Mobilising Design

This book brings together research working at the boundary between design knowledges and mobilities, offering a novel collection for both theorists and practitioners. Drawing upon detailed case studies, it demonstrates the diverse roles of design in shaping mobility at different spaces and scales: across cities; within different types of buildings and infrastructures; and through commuting, work and leisure activities.

A range of international scholars illustrate the designed mobilities of car parks, traffic lights, street benches, pedestrian wayfinding systems and accessible design in the urban environment; they examine spaces within hospitals, airports and train stations and investigate design practices for bicycles, future urban vehicles and MotoGP motorcycle racing. Other contributions explore overlooked mobile artefacts such as television and video game remote controls, 3D printing and the types of packaging which enable objects themselves to move around. This book demonstrates how the tools, assumptions and processes of design shape spaces of mobility, and also illuminates how shifts in the fluidity and circulation of people, practices and materials in turn reconfigure practices of design.

Mobilising Design develops multi-disciplinary understandings of design, drawing upon diverse literatures including design history, product design, architecture and cultural geography. By highlighting often invisible artefacts and associated knowledges and controversies, the book foregrounds the taken-for-granted ways in which everyday mobility is designed. It will be of interest to scholars in geography, sociology, economic history, architecture, design and urban theory.

Justin Spinney is Lecturer in Human Geography at Cardiff University.

Suzanne Reimer is Associate Professor in Human Geography at the University of Southampton.

Philip Pinch is Senior Lecturer in the Division of Urban, Environment and Leisure Studies, London South Bank University.

Routledge Studies in Human Geography

This series provides a forum for innovative, vibrant, and critical debate within Human Geography. Titles will reflect the wealth of research which is taking place in this diverse and ever-expanding field. Contributions will be drawn from the main sub-disciplines and from innovative areas of work which have no particular sub-disciplinary allegiances.

For a full list of titles in this series, please visit www.routledge.com/series/SE0514

Mobilising Design

Edited by
**Justin Spinney, Suzanne Reimer
and Philip Pinch**

Routledge
Taylor & Francis Group

LONDON AND NEW YORK

First published 2017 by Routledge

2 Park Square, Milton Park, Abingdon, Oxfordshire OX14 4RN

52 Vanderbilt Avenue, New York, NY 10017

Routledge is an imprint of the Taylor & Francis Group, an informa business

First issued in paperback 2018

British Library Cataloguing-in-Publication Data
A catalogue record for this book is available from the British Library

Library of Congress Cataloging in Publication Data
Names: Spinney, Justin, 1973– editor. | Reimer, Suzanne, editor. | Pinch, Philip, 1962– editor.
Title: Mobilising design / edited by Justin Spinney, Suzanne Reimer and Philip Pinch.
Description: Abingdon, Oxon, ; New York, NY : Routledge, 2017. |
Series: Routledge studies in human geography ; 69 | Includes bibliographical references and index.
Identifiers: LCCN 2016038442 | ISBN 9781138676374 (hardback) |
ISBN 9781315560113 (ebook)
Subjects: LCSH: Spatial behavior. | Design—Social aspects. |
Industrial design. | Architecture and society. | Movement (Philosophy) |
Cultural geography.
Classification: LCC GF95 .M63 2017 | DDC 711—dc23
LC record available at https://lccn.loc.gov/2016038442

ISBN: 978-1-138-67637-4 (hbk)
ISBN: 978-0-367-13872-1 (pbk)

Typeset in Times New Roman
by Book Now Ltd, London

To John Urry, whose intellectual legacy and inspiration continues to motivate our work

Contents

Illustrations

Figures

Tables

Contributors

Margo Annemans is a postdoctoral researcher in architecture at the University of Leuven/KU Leuven. Her doctoral research, conducted in collaboration with osar architects, focused on the spatial experiences of patients on their journey through the hospital and emphasised the potential for this work to provide insights into the architectural design process. Recent publications have highlighted the importance of the patient experience to building design.

Thomas Birtchnell is a Senior Lecturer in the School of Geography and Sustainable Communities, University of Wollongong, New South Wales, Australia. His interests focus on the mobilities of knowledge and innovation, cargo and containerisation, and urban futures. He has a research specialisation in South Asia. His latest book is *A new industrial future: 3D printing and the reconfiguring of production, distribution, and consumption* (2016) co-authored with John Urry.

Spencer Clark currently works at Transport for London (TfL) as a Principal Growth Area officer, helping to plan the capital's future surface transport schemes that will underpin London's growth. He played a central role in TfL's Legible London pedestrian wayfinding programme from its earliest conception stages through to implementation on street and ongoing product development. He maintains a particular interest in monitoring and evaluation of walking schemes and through this role assisted with recent research into user perceptions and use of Legible London.

Simon Cook is a PhD researcher based in the Department of Geography at Royal Holloway, University of London. His research concerns the everyday practices of everyday life; the ways in which they happen, in which they change, and what they can tell us about societies and spaces. Currently this interest is manifest in a project exploring the practice of running and his doctoral research explores the emergence and potential of run-commuting as an urban mode of mobility.

Peter Cox began working on the sociology of cycling in 2004 after returning to academia, having previously worked in the bicycle trade and in intentional communities. He lectures in the department of Social and Political Science at the University of Chester in the UK. He is author of *Moving people* (2010), coeditor of *Cycling and society* (2007) with Dave Horton and Paul Rosen; and editor of *Cycling cultures* (2015). He also serves as an advisor to the European Cycling Federation.

Martin Emanuel is historian of technology and postdoctoral fellow at Uppsala University, Sweden. He formerly held a postdoctoral fellowship at Eindhoven University of Technology, the Netherlands. Martin's expertise is at the intersection of mobility, urban, and environmental history, including also tourism history. His research focuses upon different aspects of the history of cycling culture, urban planning and traffic management.

Emily Falconer is a lecturer in Sociology at the University of Westminster, London, UK. Emily's research focuses on the politics of affect, emotion and embodied encounters in everyday life, incorporating theories from across the disciplines of sociology and human geography. Within this, Emily draws on methodologies that attempt to capture the sensual and affective experiences of everyday life and powerful moments of visceral politics.

Lino Vital García-Verdugo is a research engineer (MEng, MSc) working in the automotive industry. He also holds a PhD in Vehicle Design Research from the Royal College of Art (2012). This interdisciplinarity is reflected in experience ranging from powertrain development to semiotic analyses. His design work explores the integration of vehicle technology, architecture and media to create immersive interactive environments out of urban electric vehicles. This and his other design and academic works have been shown at renowned companies, universities and art institutions.

Ann Heylighen is research professor in the Research[x]Design group of the Department of Architecture at KU Leuven (Belgium). She leads a multidisciplinary team which studies design practices in architecture and related design domains, and explores how disabled people's spatial experience may expand prevailing ways of understanding and designing space. She has had research awards from amongst others the Research Foundation Flanders and the European Research Council.

Jayne Jeffries is a Research Associate in the School of Architecture, Planning and Landscape at Newcastle University, in the UK. Her PhD research utilised in-depth qualitative and participatory approaches to work together 'with' disabled people and their partners in the region. Her work is shaped by an underlying politics and ethics, using Participatory Action Research to develop suitable, relevant and appropriate methodologies that address change at different scales in everyday life.

Ole B. Jensen is Professor of Urban Theory at the Department of Architecture, Design and Media Technology, Aalborg University (Denmark). He is deputy director, co-founder and board member at the *Center for Mobilities and Urban Studies* (C-MUS), and Director of the research cluster in '*Mobility and Tracking Technology*' (MoTT). His main research interests are within Urban Mobilities, Mobilities Design, and Networked Technologies. He is co-author of *Making European space. mobility, power and territorial identity* (2004), with Tim Richardson; author of *Staging mobilities* (2013) and *Designing* *mobilities* (2014); the Editor of the four-volume collection *Mobilities* (2015); and co-author (with Ditte Bendix Lanng) of *Urban mobilities design: urban designs for mobile situations,* (forthcoming).

Guy Julier is the University of Brighton/Victoria and Albert Museum Principal Research Fellow in Contemporary Design and Professor of Design Culture. His research sits at the meeting point of design and the social sciences, both in terms of its contemporary practice and historical enquiry. Most recently he has developed work on the role of design in neoliberalisation processes in a monograph entitled *Economies of design* (forthcoming 2017). He is the author of *The culture of design* (third edition 2014) and co-editor of *Design and creativity: policy, management and practice* (2009).

Kim Kullman is a researcher in the Department of Sociology, Goldsmiths College, University of London. His work falls within social and cultural geography, and explores the crafting of inclusive civic spaces, practices of urban mobility and methodological crossovers between design and social science. He has published in various journals and volumes, and is the co-editor of *Care and design: bodies, buildings, cities* (2016).

Ditte Bendix Lanng is Assistant Professor in Urban Design and Mobilities at the Department of Architecture, Design and Media Technology, Aalborg University, Denmark. Ditte's research interests are within Urban Design and Mobilities Design—a combination of fields that she researches through cross-disciplinary theory, design experiments as modes of sociocultural enquiry, ethnography, and as a collective enterprise with planning and architectural practice. She has a commitment to merging relational approaches with Urban Design, and develop Urban Design's and Mobilities Design's theoretical and methodological stream of materialities as networked and active hybrids. She is the co-author (with Ole B. Jensen) of *Urban mobilities design: urban designs for mobile situations* (forthcoming).

Craig Martin is Senior Lecturer in Design Cultures, in the School of Design at The University of Edinburgh, Scotland. His research interests are primarily focused on the relationship between design and social and cultural geography, with a particular emphasis on design mobilities. Recent research projects include 'The Material Cultures of Illicit Design' and 'Ad Hoc 'Design' and Place-Making'. He convenes the DISIGN research group at The University of Edinburgh, and has published in a wide range of academic journals and edited collections including *Cargomobilities* (2015). His most recent book is *Shipping container* (2016).

Lesley Murray is a Principal Lecturer in Social Science at the University of Brighton, UK, where her research centres on urban mobilities. She previously worked as a transport researcher for the London Research Centre and the Greater London Authority, before moving to Transport for London as a Transport Planner in Strategy and Policy. Lesley's transdisciplinary research includes collaborations with artists, architects, and creative writers. She has co-edited a number of collections on mobilities: *Mobile methodologies* (2010); *Researching and representing mobilities* (2014); and *Intergenerational mobilities* (2016).

Anna Nikolaeva is a postdoctoral researcher in the project 'Smart Cycling Futures' at the University of Amsterdam and Utrecht University, the Netherlands. She previously worked at Royal Holloway, University of London, UK, as part of a research team investigating global futures of mobility. She has written about airports, mobility and public space for academic and general audiences. In her research she investigates how mobilities are governed, planned, controlled, imagined and contested in spaces of transit, cities, states and transnationally.

Philip Pinch is Senior Lecturer in the Division of Urban, Environment and Leisure Studies at London South Bank University, UK. He has recently been involved in ESPON sponsored research projects examining aspects spatial planning and territorial cohesion across the European Union. Other research interests include landscape strategies for minerals planning and post-mining environments; urban rivers and water space planning; and moto-mobilities. Recent work has explored the lessons of the UK government's wartime Utility furniture scheme for contemporary concerns about equitable and sustainable production.

Suzanne Reimer is Associate Professor in Human Geography at the University of Southampton, UK. She is interested in aspects of design, creativity, knowledge and innovation, including the gendering of creativity and design labour; and in the role of design in commodity networks. Suzanne has on-going interests in the furniture industry; modernism and design; and moto-mobilities.

Susan Robertson is a member of the Mobilities Research Group at the University of Brighton and works collaboratively across disciplines of spatial design, art, cultural geography and social science. Based on industry practice as an architect, her research investigates the modernist concerns with light, form, material and metaphor. She is particularly interested in questions of urban mobility and kinesthetics.

Justin Spinney is Lecturer in Human Geography at Cardiff University, Wales, UK. His research and teaching interests focus on the intersections between mobility, place and design, underpinned by an emphasis on the production and maintenance of power and inequality and the application of both political-economic and post-structuralist perspectives.

John Urry (1946–2016) served as director of Lancaster University's Centre for Mobilities Research (CeMoRe), before becoming a founding co-director of the Institute for Social Futures. His research interests spanned social theory, leisure and tourism; and his influential scholarship shaped debates across the social sciences. He was at the forefront of the 'new mobilities paradigm' and was a founding member of the journal *Mobilities*. Recent books include *Mobilities* (2007), *Climate change and society* (2011), *Societies beyond oil* (2013) and *Offshoring* (2014). His last book, *What is the future?*, was published in September 2016.

Chantal Van Audenhove is a professor at the KU Leuven, Academic Centre for General Practice (ACGP), Department of Public Health. She is a clinical psychologist and director of LUCAS, the Interfaculty Centre for Care Research and Consultancy of the KU Leuven. Her research primarily focuses on mental health and care for people suffering from dementia.

Hilde Vermolen is a partner at osar architects, in Antwerp, Belgium. She has been responsible for large-scale projects in the health care sector, mainly hospitals. Hilde studied architecture at the Luca, School of Arts, Brussels (Belgium), followed by an additional master's degree in Monuments and Landscape Architecture.

Justin Westgate is a design practitioner and doctoral candidate at the Australian Centre for Cultural Environmental Research, University of Wollongong, Australia. He holds a Master's degree in geography from the University of Auckland, NZ, and an honours degree in design from AUT University, NZ. His doctoral thesis investigates experimental geographies of the Anthropocene, tracing experience and thinking, with particular interest in ontological ruptures.

Simon Wind is assistant professor with a special focus on mobile ethnographies and digital technologies at the Department of Architecture, Design and Media Technology, Aalborg University (Denmark). Simon's research interests lie in the intersection of urban design, mobilities studies and (smart) technologies. Simon is a board member of the Center for Mobilities and Urban Studies (C-MUS).

Peter Wright is Professor of Social Computing in the School of Computing Science at Newcastle University, UK. His current research is concerned with the human-centred design of ubiquitous technologies. This has involved both theoretical and practical research with a focus on the experience-based and participatory design of products, services and platforms for individuals, families, communities and for organisations such as the health service.

Acknowledgements

The author and publishers would like to thank the following for granting permission to reproduce images in this work:

- Transport for London for Figures 3.1, 3.2 and 3.3;
- Deutsches Museum (Munich) Archive for Figure 4.1;
- Moulton Bicycle Company for Figure 4.2;
- Stockholm's Urban Planning Department for Figure 8.1;
- Spencer Clark for Figure 9.1;
- Scott Morrissey for Figure 9.2;
- Yoshihiko Kawauchi for Figure 10.1;
- Taylor and Francis for Figure 11.1.

Every effort has been made to contact copyright holders for their permission to reprint material in this book. The publishers would be grateful to hear from any copyright holder who is not here acknowledged and will undertake to rectify any errors or omissions in future editions of this book.

Introduction

Justin Spinney, Suzanne Reimer and Philip Pinch

The central contribution of this edited collection is to foreground relationships between individual designed objects and wider systems of mobility. In bringing a design perspective to bear on understandings of mobilities, we seek to better understand the tools, assumptions and processes through which the subjects, practices and spaces of mobility are 'made up'. At the same time, bringing a mobilities perspective (that mobility is meaningful and powerful) to design enables us to understand how shifts in the fluidity and circulation of people, practices and materials are reshaping design practices. In broad terms, *Mobilising design* emphasises the role of design as process, practice and outcome in producing mobility. Together, the book's contributors develop multi-disciplinary understandings of design, drawing upon diverse literatures including design history, product design, architecture and cultural geography.

It is important to emphasise that design is both verb and noun—process and product (Lawson 2006, 3)—something that we both do and experience as an achievement. Design is about connections: "Actors are such because they interact, shaping relations and being shaped by relations. They gain their identity in the disputes" (Yaneva 2012, 92). Here Yaneva emphasises not simply the role of the material results of design in shaping relations, but the process of design as a key way in which relations are shaped. Accordingly design should not be seen as a "cold domain of material relations" but as a "type of connector" (Yaneva 2009, 273). To understand design processes is of fundamental importance because of the ways in which they act to afford or limit the potential for transformation and alternative visions of the world. Design is a process through which objects are given human intentionality, and thus become moral and political actants.

A key influence on the insights presented in *Mobilising design* has been the substantial body of work dealing with the relations between bodies, environments and designed artefacts, particularly from Science and Technology Studies (STS) scholars and technological anthropologists (Akrich 1992; Ingold 2000, Latour 2008; Michael 2000a, 2000b). We see a need to explore the design process in terms of how artefacts come into being as much as we look at the impacts of design. As Akrich (1992, 222) has suggested:

[T]echnical objects have political strength. They may change social relations, but they also stabilise, naturalise, depoliticise and translate these into other media. After the event, the processes involved in building up technical objects are concealed …. There was, or so it seems, never any possibility that it could have been otherwise.

We also of course owe a significant intellectual debt to the insights of the 'mobilities paradigm' and in particular the ground-breaking and influential work of John Urry. Ten years ago, Sheller and Urry (2006, 208) argued that the social sciences had been 'a-mobile'—that they ignored the importance of mobility:

> Travel has been for the social sciences seen as a black box, a neutral set of technologies and processes predominantly permitting forms of economic, social, and political life that are seen as explained in terms of other, more causally powerful processes.

A primary goal in the development of mobilities scholarship was to study mobility as a force actively assembling societies and subjectivities in particular ways. Particularly important to this ongoing work was Urry's articulation of "automobility" (2000, 2007): a "complex amalgam of interlocking machines, social practices and ways of dwelling" (Sheller and Urry 2000, 739). The system of automobility was seen to have profound consequences for the spatio-temporal configurations, restructurings and practices of peoples' everyday lives.

However, despite the substantial breadth of ongoing mobilities research (Cresswell 2010b; 2012; 2014; Cresswell and Merriman 2011; Adey *et al.* 2014), specific understandings of design as both artefact and process have been underexplored. As Merriman (2015, 88) has suggested, there is much more to be written on "examining the history of the design, planning, engineering, labouring, consumption and use" of mobility spaces and artefacts. This is not to say that there has been an absence of engagement between design and mobility. For example, the object of the car—to which the automobility literature was inevitably drawn—became the starting point for examinations of driving in cities (Thrift 2004) and on motorways (Merriman 2004); accounts of the feelings and pleasures cars evoke more generally (Miller 2001; Sheller 2004); and examinations of the car's importance to concepts of nationhood and national identity (Edensor 2004, Koshar 2004). Further, Urry's (2000, 14) conceptualisation of the 'car-driver' as a complex hybrid of subject and object became centrally important to many subsequent discussions (e.g. Dant 2004; Urry 2004; Merriman 2006). Merriman (2006) also has highlighted the distributed nature of design in the context of a discussion of the car wing mirror. An ensemble of police officers, research scientists and politicians, amongst others, are seen to contribute to the design of the car with the purpose of governing the conduct of the mobile subject.

Further understandings of the design/mobility interface are provided in Levinson's (2006) work, which traces some of the regulatory and design agency decisions leading to the eventual standardisation of the shipping container. In particular, Levinson (2006) highlights relations between different manufacturers, government agencies, international standards, and shipping and haulage firms in producing the shipping container we know today. Similarly, Martin (2016) discusses the design of the container as a discrete object, but also seeks to understand the ways in which it is mobilised for other purposes, charting its shift from distant to more familiar object through redesign and subsequent entanglement in more everyday cultural ventures.

Finally, there have been some engagements with design which have taken a close focus on material spaces 'of' mobility; and have sought to understand how these designed spaces shape mobility practices. Bringing together work in mobilities, urban design and urban studies, Jensen has made an important contribution in highlighting how mobilities are staged "by means of very specific design solutions and choices" (2014, 15–16).

The distinctiveness of our focus derives from moving beyond the broader environments that shape mobilities to centrally register the importance of the design process itself. The book thus explores how artefacts become rather than importing them into accounts of mobility as prefigured. For us the emphasis is less on how artefacts are experienced, and more on how the mobility of design professionals and standards shapes material outcomes; and how arrangements of artefacts produce particular mobile practices, subjectivities and identities. We need to think in terms of tracing the emergence of roles and artefacts "as associations are formed, roles are defined, and divisions are established" (Murdoch, 1997, 744).

Structure of the book

As mobilities scholars we are concerned to explain how mobilities come to take the forms that they do, and the ways these inform and impact upon individual identities, socio-spatial relations and global processes. Providing a satisfactory explanation of mobilities design requires an understanding of the different actors and processes that produce mobilities, the assumptions they make, and the values and qualities they prioritise. The book journeys through a range of objects, materials and processes to explore the role of designers, consumer-facing business functions (such as consumer experience and marketing), and the regulatory, legal and normative frameworks that circumscribe these functions in co-producing relationships between design and mobility. By highlighting often invisible artefacts and associated knowledges and controversies our contributors foreground the taken for granted ways in which everyday mobility is designed. The chapters are grouped into three inter-related themes relating to the production of subjects and practices; the mobility of design knowledge and practices; and the ways in which design practices and artefacts connect (or disconnect) and blur boundaries between users and practitioners.

Designing mobility: the role of design in making up mobile subjects and practices

If the social as Urry (2007) suggests is 'mobile'—then the focus becomes about how movement makes up the social, and hence how design processes make up the social in particular ways. In line with Cresswell (2014) and Spinney *et al.* (2015), and building upon much of the design mobilities work cited thus far, the chapters in Part I take seriously the work of mobile artefacts as fundamental in shaping social relations that 'make up the subject'. In certain cognate literatures such as transport geography, the status of mobility artefacts has remained unquestioned; positioned as already 'made up' and acting as neutral nodes in a given network (Ernste *et al.* 2012; Spinney 2010). However, rather than view the mobile subject as 'finished' and pre-formed because of their hybridisation with various designed artefacts and environments, chapters in this part demonstrate how designed artefacts make up the subject in less predictable ways because of the manner in which artefacts interact with bodies and spaces in unique ways. Importantly, the authors attempt to excavate and foreground the logics behind the design of both spaces and objects to understand how these produce particular mobile practices and subjectivities.

Working against a broad historical canvas, Guy Julier makes a case for a series of design artefacts that typically lie outside the purview of mobilities studies as helping to usher in a 'neoliberal sensorium'. Here the design of objects like TV remote controls and computer games intersect to discipline the movements of users toward a new set of calculative 'micro-practices'. In her study of the design of Amsterdam's Schiphol airport, Anna Nikolaeva draws upon interviews with airport architects, designers and managers to identify key actors in the design process. In doing so she demonstrates how the needs, desires and capabilities of the mobile subject are 'made up' and contested by different stakeholders in the process. Using the example of the 'Legible London' wayfinding initiative for walking in London, Spencer Clark, Philip Pinch and Suzanne Reimer show how the decisions taken in the design process attempt to accommodate walking in London as part of a 'mobility fix' to ease pressure and congestion on other areas of the transport system, but also as a form of leisurely exploration. What emerges in both cases is that the meanings of walking assumed and reproduced through the decisions of design professionals rely on the prioritisation of walking as a 'rational', efficient and purposeful form of mobility. Stepping back in time, Peter Cox provides a historical STS-informed analysis of bicycle design in 1920s Germany and 1970s Britain to illustrate how external forces can shape interpretations of cycle design. Cox shows how on the one hand the meanings of existing designs are changed by inserting them into new socio-political contexts, and on the other how the meanings of new designs are changed by inserting them in to existing socio-political context. Both cases illustrate the relations between design and broader worlds of meaning and sense-making, and the ways the meanings of a mobile practice are transformed. Simon Cook demonstrates how variations of running as a mobile practice are produced through the intersection of different design elements in a train station. In particular he demonstrates the tensions between figurative design cues that attempt to eliminate running, material artefacts such as clothes and shoes that do not afford running, and informational and material cues that encourage running by both increasing and decreasing the information users have regarding the achievement of their goal (boarding a particular train). Finally, Thomas Birtchnell, John Urry and Justin Westgate's chapter indicates how the movement of design technologies from experts to lay individuals is situated within discourses of empowerment, education and democratisation. Emergent meanings of the 3D printer as an educative and democratising tool are seen to arise precisely because its movement represents a dispersal of production facilities and design practice.

Mobilising design: the mobility of design knowledge and practice

John Urry's (2000) landmark mobilities text *Sociology Beyond Societies* emphasises that "sociality and identity are produced through networks of people, ideas and things moving rather than the inhabitation of a shared space such as a region or nation state" (Cresswell 2010b, 551). That is, interconnected movements profoundly reshape what we do and how we do it. Within this premise of increased circulation and fluidity, Urry asks us to consider how "objects are materially produced and symbolically conceived of" (2000, 66). In the language of actor-network theory, rather than 'black box' design processes and encouraging them to be viewed as singular and linear entity, Urry (2000) encourages us to unpack design and focus on the ways in which movements, contingencies and multiplicities of design shape both design outcomes and practice.

The chapters in Part II typically understand the design process as an intersection and potentiality where the designed assemblage comes into being through ongoing movement and negotiation amongst multiple stakeholders. As Yaneva (2012, 544) has suggested, designed objects are the outcome of "a continuous series of negotiations, struggles and compromises between specific professional circles that have some common agendas to address but pursue different objectives". Contributors to this part demonstrate the ways in which designed objects (and thus mobilities) are constantly transformed by actors and agents including designers, security personnel, engineers, marketers, advertisers, coders, and users. Despite recognition of the increasingly dispersed and fragmented nature of design practice and production (Julier 2000; Henry *et al.* 1996), the impact of this mobility on design practices has been little explored. The chapters in this part argue for an ontology of the place of design that recognises the powerful ways in which the increasingly shifting, unmoored and dispersed nature of design practice impacts upon its outcomes. Rather than conceptualise design as one side of a producer/consumer boundary, design becomes a set of nodes in a shifting network. Here a design object or project becomes more a 'complex ecology than a static object' (Yaneva, 2012, 283). This focuses attention on the fluid, unfinished and uncertain nature of design materials, knowledges and practices, and the roles of different actors and actants in producing them as such.

Through an analysis of packaging design, Craig Martin demonstrates how the mobility of various ideas and materials becomes a core determinant of design geographies: from the circulation of ideas in the initial conceptualisation of a design solution; through the global mobilities of raw materials and finished goods; to the mediation of information in consumer culture. Martin Emanuel focuses on the (partial) transition from hand-signalled to mechanised control of urban mobility on the roads of interwar Stockholm. In analysing controversies around the early development of traffic lights, Emanuel aims to understand what lights 'do' in a Latourian sense, presenting the design of the now taken-for-granted traffic light as arising from transatlantic circulation of knowledge and technology, and from the connecting together of a diverse set of actors including Swedish technology company AGA, the Swedish Police Force, the US firm General Electric, motoring journalists and the Royal Automobile Club. Philip Pinch and Suzanne Reimer demonstrate the multiple and dislocated nature of design through the example of MotoGP, the premier motorcycle racing world championship. Design is seen to traverse a range of locations, including mechanical engineering, the design of electronic control systems, the regulation of international motorcycle racing and the physical body of the rider. In such a reading the outcomes of design practice are never settled, but rather design continually responds to different riders, racing circuits and rule changes. In a similar vein but focusing on the travels of one influential individual, Kim Kullman demonstrates how discourses of Universal Design are increasingly shaped through mobility. Kullman demonstrates how transnational relations of knowledge as ideas, models and techniques are produced and transformed by the physical travel experiences of Professor Kawauchi. Finally, through a case study of a suburban car park in Aalborg (Denmark) Ole B. Jensen, Ditte Bendix Lanng and Simon Wind illustrate the importance of context in determining the visual and practical affordances of objects and spaces. In doing so they demonstrate the ways in which uses and things are always connected into wider networks conditioning their actual and potential uses.

Design knowledges: making connections

As we have discussed, design must be understood as a distributed process; the intersection of forms of expertise and values. This highlights the need to understand design as an unstable and dispersed accomplishment which is fundamentally about making connections: design connects us (or not) to the social and physical world; to each other. Design is thus an inherently political practice even if design professionals have been at pains to obscure it. Designed artefacts are the materialisation—and indeed fixing and representation—of particular values, interests and dispositions. As such they tell a particular story, or more accurately they give us the 'ending'—the outcome. We cannot tell just by looking at something how it has come to take this shape and not that or why it enables this use but disables that use. As Yaneva suggests, "we cannot understand how a society works without appreciating how design shapes, facilitates, conditions and makes possible everyday sociality [...] design helps make the social durable" (2012, 280–281).

The chapters in this third and final part explore the ways in which the design process connects or disconnects different fields of users, professionals and forms of representation. Here the chapters highlight the importance of intersections as the borders where translations and distortions occur, where some understandings take precedence over others. Yaneva (2012, 282) argues that to tackle design as a type of connector; "as a mechanism for setting the world in motion" we need to go beyond the discourses of designers and inventors, arguing that we also need to attend to the cultures and practices of designers, and to follow "what designers and users do in their daily and routine actions". Storni (2012, 89) suggests a focus on "the process of design, in the becoming of its results and in the relation between process and outcome" to foreground the movements and transformations between product and process. Accordingly, chapters in Part III focus on the ways in which standards, regulations and other virtualised and 'invisible' actants of design intersect to shape mobility.

Using the case of the urban quadricycle, Lino Vital García-Verdugo demonstrates how a focus on vehicle design that moves beyond the aesthetics of the showroom, highlights the potential for design to connect people and environments in new and more civically progressive ways. In doing so he highlights the fact that for this to happen, car designers must connect with other cognate nodes such as urban design. Margo Annemans, Chantal Van Audenhove, Hilde Vermolen and Ann Heylighen focus on the problem of designing spaces from the view of a particular mobile subject—the hospital patient. Annemans *et al.* are concerned to evaluate which processes and media enable the best representation and translation of differently mobile subjects into design qualities and outcomes. Their findings illuminate the connections between different design outcomes and the media through which mobility are represented. Jayne Jeffries and Peter Wright bring together insights from mobilities and human–computer interaction to explore the borders that exist between 'designers' and 'users'. By charting the exchange of skills, perspectives and knowledge's between research-designers and participant-users, Jeffries and Wright chart a shift from design as a strictly bounded way of knowing and 'telling' to design as a porous connector. Emily Falconer provides an alternative reading of the train station as designed space through an affectual analysis of mobility patterns. Her analysis sheds light on how people's senses and feelings of community are constituted in relation to forms of mobility and waiting. Journeying with commuters as they traverse through the multisensory atmospheres of an early morning train station, a familiar yet silent train carriage and a noisy, neon-lit, smell-ridden and body-filled metropolitan station, Falconer reveals how the affordances of space create feelings of connection and separation, belonging and exclusion, community and individualisation. Finally, through an examination of a specific shared street space—New Road in Brighton, Lesley Murray and Susan Robertson explore the potential of drawing as design practice to interrogate the complex connections between the design and mobile practices of street space. In particular Murray and Robertson focus on the ways in which designers attempt to represent the multisensory nature of mobility and differential power relations of mobile subjects through such a 'static' medium. What arises from this is an understanding of the ways in which drawing is central to conceiving and giving meaning to different mobile practices.

Our concluding chapter identifies a range of themes that signpost both the collective contribution of chapters and avenues for further research. In particular, it highlights the importance of understanding designed objects and designers as moral and political actors. We would like to end with a note of thanks to our authors for their continuing interest and support throughout the production of the book. The volume has its origins in a special session at the 2013 Royal Geographical Society/Institute of British Geographers Annual International Conference as well as a one-day workshop, *Designing Mobilities 2015*, held at London South Bank University, 14–15 April 2015. We are grateful to all of the participants at both events for their thoughtful reflections and contributions.

References

Ackrich, M., 1992. The de-scription of technical objects. In *Shaping technology/building society*, ed. W. Bijker and J. Law. Cambridge, MA: MIT Press, pp. 205–224.

Adey, P., Bissell, D., Hannam, K, Merriman, P. and Sheller, M., eds., 2014. *The Routledge handbook of mobilities*. London: Routledge.

Cresswell, T., 2010a. Towards a politics of mobility. *Environment and Planning D: Society and Space*, 28: 17–31.

Cresswell, T., 2010b. Mobilities I: catching up. *Progress in Human Geography*, 35: 550–558.

Cresswell, T., 2012. Mobilities II: still. *Progress in Human Geography*, 36: 645–653.

Cresswell, T., 2014. Mobilities III: moving on. *Progress in Human Geography*, 38: 712–721.

Cresswell, T. and Merriman, P., 2011. *Geographies of mobilities: practices, spaces, subjects*. Farnham: Ashgate Publishing.

Dant, T., 2004. The driver-car. *Theory, Culture and Society*, 21: 61–79.

Edensor, T., 2004. Automobility and national identity: representation, geography and driving practice. *Theory, Culture and Society*, 21: 101–120.

Ernste, H., Martens, K. and Schapendonk, J., 2012. The design, experience and justice of mobility. *Tijdschrift voor Economische en Sociale Geografie*, 103: 509–515.

Henry, N., Pinch, S. and Russell, S., 1996. In pole position? Untraded interdependencies, new industrial spaces and the British motor sport industry. *Area*, 28: 25–36.

Ingold, T., 2000. *Perception of the environment: essays in livelihood, dwelling and skill*. London: Routledge.

Jensen, O.B., 2014. *Designing mobilities*. Aalborg: Aalborg Universitetsforlag.

Julier, G., 2000. *The culture of design*. London: Sage.

Koshar, R., 2004. Cars and nations: Anglo-German perspectives on automobility between the world wars. *Theory, Culture and Society*, 21: 121–144.

Latour, B., 2008. A cautious Prometheus? A few steps toward a philosophy of design (with special attention to Peter Sloterdijk). In *Proceedings of the 2008 Annual International Conference of the Design History Society*, Falmouth, 3–6 September 2008. Ed. F. Hackne, J. Glynne and V. Minto. e-Books, Universal Publishers, pp. 2–10. Available at: http://www.bruno-latour.fr/article?page=3, accessed 10/10/2016.

Lawson, B., 2006. *How designers think: the design process demystified*. London: Routledge.

Levinson, M., 2006. *The box: how the shipping container made the world smaller and the world economy bigger*. Princeton, NJ: Princeton University Press.

Martin, C., 2016. *Shipping container*. London: Bloomsbury.

Merriman, P., 2004. Driving places: Mark Augé, non-places and the geographies of England's M1 motorway. *Theory, Culture and Society*, 21: 145–167.

Merriman, P., 2006. 'Mirror, signal, manoeuvre': assembling and governing the motorway driver in late 1950s Britain. *The Sociological Review*, 54: 75–92.

Merriman, P., 2015. Mobilities I: Departures. *Progress in Human Geography*, 39: 87–95.

Michael, M., 2000a These boots are made for walking . . .: mundane technology, the body and human-environment relations. *Body and Society*, 6: 107–126.

Michael, M., 2000b. *Reconnecting culture, technology and nature*. London: Routledge.

Miller, D., ed., 2001. *Car cultures*. Oxford: Berg.

Murdoch, J., 1997. Inhuman/nonhuman/human: actor-network theory and the prospects for a nondualistic and symmetrical perspective on nature and society. *Environment and Planning D: Society and Space*, 15: 731–756.

Sheller, M., 2004. Automotive emotions: feeling the car. *Theory, Culture and Society*, 21: 221–242.

Sheller, M. and Urry, J., 2000. The city and the car. *International Journal of Urban and Regional Research*, 24: 737–757.

Sheller, M. and Urry, J., 2006. The new mobilities paradigm. *Environment and Planning A*, 38: 207–226.

Spinney, J., 2010. Performing resistance? Re-reading practices of urban cycling on London's South Bank. *Environment and Planning A*, 42: 2914–2937.

Spinney, J., Aldred, R. and Brown, K., 2015. Geographies of citizenship and everyday (im) mobility. *Geoforum*, 64: 325–332.

Storni, C., 2012. Unpacking design practices: the notion of thing in the making of artifacts. *Science, Technology & Human Values*, 37: 88–123.

Thrift, N., 2004. Driving in the city. *Theory, Culture and Society*, 21: 41–59.

Urry, J., 2000. *Sociology beyond societies*. London: Routledge.

Urry, J., 2004. The 'system' of automobility. *Theory, Culture and Society*, 21: 25–39.

Urry, J., 2007. *Mobilities*. Cambridge: Polity.

Yaneva, A., 2009. Making the social hold: Towards an actor-network theory of design. *Design and Culture*, 1: 273–288.

Yaneva, A., 2012. *Mapping controversies in architecture*. London: Routledge.

Part I

Designing mobility

Mobile subjects and practices

1 From the movement *of* things to movement *in* things

Object-environments and the neoliberal sensorium

Guy Julier

Introduction

If there was ever a watershed moment for neoliberal economic practices, it was 1985–6. Reagan in the USA and Thatcher in the UK were by then comfortably into their respective second terms, overseeing a set of deregulations of transport, shipping and finance that resulted in more rapid and more extensive ways by which goods, people and money could be moved. On 27 October 1986, the so-called Big Bang of the London Stock Market occurred. Thenceforth trading commissions were no longer fixed, overseas investors could operate on the London markets, the roles of broker and market maker were merged and deals could take place by telephone and computer rather than face-to-face (Clemons and Weber 1990). Money could move fast through new networks at new speeds.

Alongside these changes, a number of electronic devices came into being that had roles in the unfolding of a new, neoliberal sensorium. This sensorium refers to a particular set of affective, embodied dispositions and behaviours through which neoliberal structures of being are played out. In particular, the sensorium involves small-scale gestures that are connected to extra-bodily movement and, in turn, provide feedback in highly calculated forms. Crudely speaking, there is a new connection of gesture to financial movement in the actions of stockbrokers following the 1986 Big Bang. Reading and interpreting data on screens leads to decisions regarding the allocation of finance into global flows that are carried out through movements that involve inputting commands, which result in further data feedback on-screen. But we find a new regime of gesture, calculation and feedback in connection to other devices that are more germane to wider populations, although they still exist within this frame of the neoliberal sensorium.

This chapter pays attention to how specific design artefacts that typically lie outside the purview of mobilities studies are nonetheless active in the shaping of specific expectations, understandings and practices of mobilities. It aligns this consideration with the emergence of a wider framework of neoliberal subjectivity. In effect, this involves considering a kind of microbiopolitics of mobilities at work in which design is active (Foucault 2008; Thrift 2008; Väliaho 2014). Finally, the chapter reflects on how this latter disciplining of movement continues to function and what implications it has for thinking about mobilities and design.

A key design typology here is what I call the 'object-environment', which shares some features with Thrift's (2004) concept of 'movement-space'. This takes the discussion of design and mobilities beyond 'objects that move' or 'environments in which you move' to think about the role of devices as interfaces that deliver a spatialised set of relations through which the user negotiates their progress. Interaction in and movement through these spaces are invariably calculable. At the same time, this is the result of highly embodied actions that are very specific to the kinds of devices under consideration. Looping back, it is not coincidental that early versions of these devices and an emergent discourse on the interface would appear in this watershed moment of neoliberalism and new conceptions of mobility in the mid-1980s.

Design discourse, mobilities and movement: four frameworks

There have been three dominant notions of movement at work in design culture studies and design history. It is safer to talk of 'movement' here, rather than 'mobilities', for to date, design history and design culture studies have neither explicitly nor consistently engaged with the mobilities turn at work in the social sciences since the mid-2000s. Certainly, some design historians have written about forms of transport and its devices. These include work on cruise shipping (Quartermaine and Peter 2006), the London underground rail network (Forty 1986; Lawrence 2008) and railway timetables (Esbester 2009). However, these studies have only implicitly addressed or challenged how we might conceive of mobilities as presented in this book. So it is more productive to take a side step into a wider view of how the movement of design artefacts has, to date, been conceptualised in design history and design culture studies. This takes us closer to a fourth framework that I want to propose in order to engage more directly with mobilities concerns.

The first framework of movement in design discourse is perhaps the most established. This is concerned with the movement of goods, and unfolds a moral argument as to what is 'good enough' to be moved. It has its origins in the later-nineteenth century through the design reform movement spearheaded by such figures as Henry Cole, John Ruskin, William Morris and Christopher Dresser. In this framework, design (or, otherwise, the exercise of good taste) was to present an ethical challenge that harnessed sensitivity and control as against the rampant commercialism of modern production and consumer culture (Dutta 2009). This stemmed from what had been experienced through international expositions such as the 1851 Great Exhibition and the growth of global trade. Movement, in this discourse, was taken to involve the importation and exportation of commodities, in part motivated by concerns around trade balance, but also in maintaining discernment and rigour in terms of what might be deemed acceptable. Thus the concept of 'good design' that pervaded discourse through most of the twentieth century was largely wrapped around the design quality of industrial products and their passage from producer to consumer.

The second framework emerged just as this discourse of good design began to recede. In academic work this is where cultural studies began to impact on design history and design studies in the 1980s. Here, the possibility that objects move through various 'moments' of being as they pass through domains of production, mediation and consumption (Lees-Maffei 2009) comes into view. Hebdige's (1988) essay on the Vespa motor-scooter, in which he skilfully rides through a series of 'moments' in its biography, is a seminal piece here. Hebdige (1988) and others (e.g. Johnson 1986; du Gay *et al.* 1997), accept that the design object is mutable as it moves through different moments. This mutability is both in the meanings that the object is accorded and in what that object *is*: it is a device here, an image there, an aspiration or rumour in other places. Movement may thus be better termed as 'circulation' here; and design artefacts are taken to circulate through different cultural registers and formats.

The third framework is where design itself provides the environment for movement. Here, drifting, browsing, looking or wandering through come to the fore. The experience of shopping is a visual one and, with this, considerations of the role of photography in the construction of certain modes of visuality, the configuration of retail spaces or the construction of tourism and leisure environments become important (Buck-Morss 1991; Urry 1990, 1995). Of interest here is the design of spaces of consumption in such a way as to afford some freedom of movement and a measure of consumer sovereignty within highly scripted and regulated environments (Miles 1998). It is not coincidental that much of the literature that emerged here—stemming from a mix of critical geography, anthropology, visual culture studies and design studies—appeared in the 1990s, the decade that also gave us the world-wide web and that also saw further sophistication in video games. The promise of multi-linear movement through and immersion in spaces was not just confined to shopping malls or theme parks but also appeared in digital formats in this decade.

A fourth framework of movement that I am presenting for consideration combines these while also adding a typology of the 'object-environment'. Here, we find objects that act as interfaces between the user and various physical and virtual spaces to produce immersive and yet calculable experiences. The user moves, via specifically designed interfaces, between spaces in ways that capture their psychic and bodily attention. And yet, in this, users do not become 'lost' in (virtual) space. Rather, progress is subject to continual, quantitative feedback.

This fourth framework implies new forms of movement and object relations that propose a different appreciation of mobility from the other three. The first aligned the movement of things with moral debates concerning mass manufacture and taste. The second saw movement of things as being their transmutation through different contexts as well as forms of representation, producing different meanings. The third focused on movement within environments and the supposed freedoms and affects of this. The fourth framework that I'm adding here has less to do with the first, but to some degree folds together the second and third. It is about transformations and translations between mediums while promising multiple pathways within prescribed settings. Additionally, though, it invariably includes continuous systems of quantitative feedback that articulate progress but also provides structures for anticipating future movements. Coupled with this, we find this regime of calculation being connected back to micro-gestural, embodied practices.

Close analysis of particular object-environments helps to explore this fourth framework for wider thinking concerning design and mobilities. By building through the different scales in which these object-environments functioned we can begin to appreciate the significance of their design in resetting the coordinates and vectors of everyday life and mobilities.

New object-environments of the mid-1980s

In 1985, the consumer electronics company Philips released the Magnavox universal remote control, the first device that could interact with both television and video-cassette recorders. The TV remote was to be found in 27 per cent of US households in 1985 and by 1993 it was in 90 per cent of them (Bellamy and Walker 1996). Hitherto, TV viewing largely involved coordination between printed schedule listings and the television's buttons; the heavy trudge across the lounge to switch either the television set or the video recorder on or off; or putting up with a show only because the latter task was too onerous at the end of a day's work. Now, from the comfort of their sofa, the viewer could cruise between channels, browse what was on offer, switch stations during commercial breaks, record programmes or even sections of them, playback, rewind or fast-forward the tape.

In 1986, the Nintendo Entertainment System (NES) was introduced in Europe, having been tentatively rolled out in the USA in 1985 and in Japan in 1983. The NES included a more advanced control that was relatively more ergonomic, comfortable and intuitive compared with its forerunners (Kohler 2004). 'Second generation' (1976–83) consoles from manufacturers such as Atari, Arcadia or Vtech involved clunky joysticks and separate number pads that often demanded that you took your eyes off the screen. The NES console could be held in both hands and had direction, select and 'fire' buttons sitting neatly in a row. It was bundled with Super Mario Bros which became a significant entry-level video-game for consumers. Now, gamers would sway, jump a little or flinch as they worked their way through different levels, engrossed in the picture, submerged in the Super Mario musical anthem while progress, attainment and time were scored against each other.

Immersive yet calculable; stationary yet moving; ocular yet embodied: such objects as the NES games console and the Magnavox remote control (and many others of their era, as we shall see) enabled new formats and expectations of movement. Playing Super Mario Bros proposed the notion of episodic measurement in everyday home entertainment. Players made their way through successive stages, each level rated with a score. Each action of the avatar requires skilful manipulation of the console and deep concentration. Moving through obstacles, carrying out tasks, making it to the end of the game and the scoring of progress involved intensified movement.

The Magnavox TV remote and the NES video-games console may be best described as object-environments. While only minimally spatial in themselves, they both opened onto more extensive virtual environments. The TV remote reconfigured the relationship between TV viewer, TV room and television. In one way, viewers find themselves more separate from the green-eyed monster in the corner of the living-room. They no longer had to physically approach the television so frequently. On the other, the interactivity that was afforded took the viewer further into broadcasting worlds beyond it. Zapping, for example, involves moving between stations, and thus registers in continuous flow (Metz 1974) rather than, perhaps, watching discretely planned programmes. The world of television and the video-cassette recorder provides an open architecture of experiences, of time-frames made more accessible via the universal remote control.

We also see a conceptual shift in the collapsing of object and image. By this I don't just mean the relationship of the console to the screen image that resulted in underlining ocular work as a wholly embodied activity. The spatial and temporal gap between bodily action and the thing on screen was closed down; the flat screen and the games consoles became interfaces. In playing Super Mario Bros or in switching through TV stations and video playback, the resulting somatic memory—the repeated bodily actions that get embedded into neurological pathways—was associated with achievement that is spatialised through a series of on-screen images (Ash 2012, 11).

These new object-environments of the television remote and the video-games console blur boundaries between stillness and movement, objects and environments, things and representations. The following sections of the chapter reflect upon how the emergence of this material, visual and spatial cultural phenomenon also opens onto a new experience of mobility that was consolidated in the mid-1980s and has been with us since.

Other object-environments

These mid-1980s objects were not exceptions. Traces of this kind of blurring between object and environment, the introduction of movement therein and the intensification of embodied action through closing the input/output gap can be found in many other important developments of the time. For example, the introduction in 1985 of the PostScript computer programming language by Adobe Systems meant that, in computer graphics, text and image could be combined on one page, on-screen. For the first time, computers could present 'What You See Is What You Get' which could then be extended into the printed output. This was wrapped around the appearance of Apple personal, desktop computers such as the Macintosh 128k (1984), the Macintosh Plus (1986), the Macintosh SE (1987) and the Mac II (1987) (Pfiffner 2002). Like the TV remote, new, more deeply entangled relationships between hardware and software were forged. The bodily actions of designing a page layout on-screen and its outputting to print were brought closer together. The idea of swift transmutation between electronic and material matter was not necessarily new—this had been achieved, for example, through the telegraph in the early nineteenth century. However, Adobe Systems and Apple made this concept accessible and personalisable for a wider audience.

If we are to take this discussion closer to traditional transport design, then it is noteworthy that London Transport introduced its Travelcard system in 1983. This removed the need to buy separate tickets for individual tube or bus journeys. The Travelcard was extended to overground rail services—via the Capitalcard— in 1985. Whilst not the first integrated and unlimited-use ticket—Paris's Carte Orange achieved this in 1975—the Capitalcard had implications for how mobility might be conceived. By facilitating seamless movement through ticket hall to train, or from tube station onto bus, the Capitalcard implied a greater sense of sovereignty on the part of the traveller. Just as the TV remote allowed switching between channels or between live television and video playback, so the Capitalcard simplified transfers between transport modes. As such, travel would sometimes become more like 'quickened drift' in a network. Having paid a flat rate for unlimited use within a specified transport zone, users had less to lose in experimenting with routes home, for instance. Mobility, here, implied a different sense of directionality and progress—perhaps more multi-linear, perhaps more immediate. For Londoners, and by extension to other cities, the vectors of everyday life were being reset through the possession of a simple card.

The interfaces of the newly digitised London stock market also adjusted the relationships between user, information and space. On-screen trading meant that money could be moved more quickly, more easily and further as spatial and temporal relationships between brokers were reconfigured. While financial markets involve rapid movement of money between nodes, it is as well to remember that those nodes themselves and the hardware that joins them constitute their own material culture, made up of telephones, screens, cabling, office furniture and so on (Mackenzie 2008, 2014).

Within this, stockbrokers became active in a different kind of 'timeworld' where temporal coordination between other stockbroking nodes is vital for their functioning (Knorr Cetina 2003). The screens that came into use are 'epistemic things': they show information but also do some of the 'thinking'; they carry knowledge but also do some of the work of interpreting that knowledge through, for instance, calculating stock trends and showing graphs (Knorr Cetina and Bruegger 2000). The object-environment of the trader's offices are 'global in scope' but 'microstructured in character' (Knorr Cetina 2001). Their hard surfaces and glowing screens open onto a world of exchanges, measurements and expectations. These add up to the embedding of anticipatory practices that attempt to rationalise and pre-empt future events within a logic of calculation (Ash 2010, 661; Stiegler 2010, 18; Ash 2012, 7–9). Like the video-game player, the stockbroker is relatively immobile but their actions reach into other contexts of mobility.

These examples extend the discussion as to what particular object-environments were doing in the mid-1980s. Just as the ways by which finance was handled were changing, so were other bodily performances. Taken collectively, the object-environments I have discussed this far offered a certain emancipation in terms of the removal of barriers to where one might go. At the same time (and increasingly for their descendants, as in the case of the Capitalcard) movement itself is tied into on-going regimes of numeric feedback. In this, we find the roots of alignments between individual performance and a wider concept of highly audited culture and society (Power 1999; Strathern 2000).

Devices help their users achieve certain direct goals—designing a company report, getting to a station platform, trading stocks, commodities or currencies. But they are also active in producing certain expectations and understandings of spatial and temporal possibilities. They can encode particular anticipations of what mobilities might mean. The next section pursues how this is done in more detail.

Interfaces and bodies

It is not insignificant that devices such as the Magnavox TV remote, the NES, the Capitalcard or Adobe Postscript emerged in the mid-1980s alongside massive structural changes in the operation of Western economies. Such transformations included the deregulation of banking and a collapse of the separation of commercial and investment banks; and the privatisation and deregulation of much transportation and shipping in both the US and the UK. Money and people were beginning to move in strange new ways.

In drawing these new mobilities, devices and embodiments together, it is useful to review how Väliaho (2014) discusses the idea of somatic memory and digital information in relation to 'the neoliberal brain'. His study of visual-kinetic representations, their affect and how this produces certain dispositions and forecloses others brings into line, for example, the design of video-games with economic behaviours. Movement through episodic virtual spaces that are heavily quantified will place the player in a series of anticipations. Setting up the next move or collecting data that strengthens one for that are as germane to video-game playing as to everyday neoliberalism (Mirowski 2013; Beer 2015).

The ways by which interfaces provided a conduit for bodily performances was already being opened out through some academic enquiry in the mid-1980s. By the time the NES games console was for sale on the European market, Sherry Turkle had published *The Second Self* (1984) in which she challenged the division between the physical and the social. Here the term 'interaction' appears as not something that is exclusive in terms of human-to-human communication, but also happens between human and machine. In this, humans are 'rewired': their cognitive processes become adjusted to encompass the language and processes of digital technology. Humans become hybrid with these machines. In 1985, Donna Haraway published the first version of her 'Cyborg Manifesto' in which she noted the loosening of the boundaries between animal and human, animal-human and machine, and the physical and non-physical. Lucy Suchman (1987) argued that cognition was situated and contextual. This means that far from being fixed, cognition is formed in ever-changing ways according to dynamic interactions between humans and their material environments. Meanwhile, Winograd and Flores (1986) were also concerned with the functioning of the interface. Again, they saw this as being about a dynamic relationship rather than a separate realm. The user is *in* it rather than a separate viewer *of* it. In this way, users were seen to be breaking away from a sense of Cartesian dualism whereby the experience of external phenomena is assumed to largely engage the mind rather than the whole body.

Subsequently a new sensorium has emerged whose effects are still known today. Within this, the interface is a key feature in generating a particular biopolitics. Thrift (2008, 187) goes further to describe this as engaging *microbiopolitics*— small-scale actions undertaken in tiny slices of time but which are nonetheless sensed. This is where gesture becomes part of the embodied repertoire of what it is to be in the advanced industrial world. It is also where 'qualculation' becomes central to everyday practices, where a continual assessment at every encounter is rife in modern life (Cochoy 2008). This involves the qualitative assessment of phenomena through digital or analogue data (Thrift 2008, 100).

Within this new sensorium the meanings of gestures also change: 'push', 'hit', 'stroke', 'caress', 'flick', 'seize', for instance, all find new functions. These constitute the gestural language that give access to and movement through information spaces. Meanwhile, a continual data response takes place, giving feedback on progress that forms the information for further calculations on future movements. In this, devices play their part in the formation of particular expectations of mobilities. These include an expectation of the freedom to choose routes within a set menu of options. Devices also engage a disciplining of the self to regimes of calculation and accountability that sit within neoliberal, ideological structures.

Gestures associated with mobilities and qualculation have become more elaborate. In 2003, the London Travelcard was replaced by the Oyster card. Pressed against a sensor, barriers to your underground train journey open. Later, you can log into a dedicated Transport for London website, check your journeys and see how much they have cost. Tracking your travels and presenting this as experiential capital is taken to another level for 'year out' enthusiasts. Smartphone applications such as TrackMyTour and Travelog allow you to map global wanderings for friends and family to see. Jauntlet connects Facebook, Foursquare check-ins, Twitter and Instagram to produce a travel blog. Distances travelled are calculated automatically.

Conclusion

In this chapter I have attempted to focus upon particular devices themselves and to illuminate their connection back to larger contexts of mobilities. The TV remote and the video-game both engaged new forms of bodily action and play. But these intimate, domestic devices also opened onto new relations between subject and object. The new embodied skills of their use were connected to new ways to conceive of spaces. In terms of the video-game, these spaces were progressive, episodic and measured, each one existing in anticipation of the next one. Gaming involved the mastery of successive things and calculations, via tiny kinetic-spaces leading to the next one and the next one. New repertoires of gesture linked physical to virtual space and the body to numeric regimes.

The historical moment of the surfacing of the object-environments discussed in this chapter coincided with the intensification of flexible, capital accumulation. The movement of goods and people was being freed up through successive deregulations of transport and shipping in the 1980s. However, perhaps more momentous was the way in which the loosening of infrastructures for the global flow of finance was being established. The Big Bang of October 1986 instituted the possibility of easier mobility of money entirely mediated by screens. A new gestural repertoire had also to be learnt by stockbrokers. In parallel, it is useful to note that the flat image has a spatiality to it and beyond it (Ash 2010). The trading screen displays items in an architectonically planned space but these items also refer to architectures (of finance, logistics, resources, capital, for instance) outside the glowing screen.

Of course, in terms of the sheer numbers of the wider population, few people directly engaged in this kind of financial work. But it represented the sharp end of enormous cultural and economic changes that were taking place. New gestural repertoires were being learnt elsewhere, with other object-environments that, nonetheless, engaged 'stockbroker-like' forms of anticipation and calculation. In this, the object-environment is active in the formation of neoliberal sensorium. It interfaces between different realms of movement and in that transmutation, the mobile subject is recast as calculative and accountable.

The four frameworks for design and mobilities that are outlined in this chapter testify to the many different ways that are available for approaching their relationships. They show the broad range of design objects, environments and object-environments that can come into the discussion, and the variety of mobilities that can be considered. The discussion of object-environments and the 'fourth framework' presented here takes us beyond considering the physical movement of artefacts or people through space and time. Instead, some designs, and particularly interfaces, are active in shaping our understandings of mobilities, fashioning expectations of them and what they mean.

Acknowledgement

The sections 'New object-environments of the mid-1980s' and 'Interfaces and bodies' are adapted from Chapter 2 of *Economies of design* (forthcoming), published by Sage Publications.

References

Ash, J. 2010. Architectures of affect: anticipating and manipulating the event in processes of videogame design and testing. *Environment and Planning D: Society and Space*, 28: 653–671.
Ash, J. 2012. Attention, videogames and the retentional economies of affective amplification. *Theory, Culture and Society*, 29: 3–26.
Beer, D. 2015. Productive measures: Culture and measurement in the context of everyday neoliberalism. *Big Data and Society*, January-June, 1–12.
Bellamy, R and Walker, J. 1996. *Television and the remote control: grazing on a vast wasteland.* New York: Guilford Press.

Buck-Morss, S. 1991. *The dialectics of seeing: Walter Benjamin and the Arcades Project.* Cambridge, MA, MIT Press.

Clemons, E. and Weber, B. 1990. London's big bang: a case study of information technology, competitive impact, and organizational change. *Journal of Management Information Systems*, 6: 41–60.

Cochoy, F. 2008. Calculation, qualculation, calqulation: shopping cart arithmetic, equipped cognition and the clustered consumer. *Marketing Theory*, 8: 15–44.

du Gay, P., Hall, S., Janes, L., Mackay, H. and Negus, K. 1997. *Doing cultural studies: the story of the Sony Walkman.* London: Sage/The Open University.

Dutta, A. 2009. Design: on the global (r)uses of a word. *Design and Culture*, 1: 163–86.

Esbester, M. 2009. Designing time: the design and use of nineteenth-century transport timetables. *Journal of Design History*, 22: 91–113.

Forty, A. 1986. *Objects of desire: design and society since 1750.* London: Thames and Hudson.

Foucault, M. 2008. *The birth of biopolitics: lectures at the Collège de France 1978–1979.* Ed. M. Sennelart. Tr. G. Burchell. Basingstoke: Palgrave.

Haraway, D. 2006 [1985]. A Cyborg Manifesto: science, technology, and socialist-feminism in the late 20th century. In *The International Handbook of Virtual Learning Environments.* Ed. J. Weiss, J. Nolan, J. Hunsinger and P. Trifonas. Dordrecht: Springer, pp. 117–158.

Hebdige, D. 1988. *Hiding in the light: On images and things.* London: Comedia.

Johnson, R. 1986. The story so far: and further transformations. In *Introduction to contemporary cultural studies.* Ed. David Punter. London: Longman, pp. 277–313.

Kohler, C. 2004. *Power-up: how Japanese video games gave the world an extra life.* New York: BradyGames.

Knorr Cetina, K. 2001. Objectual practice. In *The practice turn in contemporary theory.* Ed. T.R. Schatzki, K. Knorr-Cetina and E. Von Savigny. London: Routledge, pp. 184–196.

Knorr Cetina, K. 2003. From pipes to scopes: the flow architecture of financial markets. *Distinktion: Scandinavian Journal of Social Theory*, 4: 7–23.

Knorr Cetina, K. and Bruegger, U. 2000. The market as an object of attachment: exploring postsocial relations in financial markets. *Canadian Journal of Sociology/Cahiers canadiens de sociologie*, 25: 141–168.

Lawrence, D. 2008. *Bright underground spaces: the London tube station architecture of Charles Holden Lawrence.* Harrow Weald: Capital Transport.

Lees-Maffei, G. 2009. The production-consumption-mediation paradigm, *Journal of Design History*, 22: 351–376.

MacKenzie, D. 2008. *Material markets: how economic agents are constructed.* Oxford: Oxford University Press.

MacKenzie, D. 2014. At Cermak. *London Review of Books*, 36(23): 25.

Metz, C. 1974. *Language and cinema.* Trans. Donna Jean Umiker-Sebeok. The Hague: Mouton.

Miles, S. 1998. The consuming paradox: a new research agenda for urban consumption. *Urban Studies*, 35: 1001–1008.

Mirowski, P. 2013. *Never let a serious crisis go to waste: how neoliberalism survived the financial meltdown.* London: Verso.

Pfiffner, P. 2002. *Inside the publishing revolution: the Adobe story.* San Jose, CA: Adobe.

Power, M. 1999. *The audit society: rituals of verification.* Oxford: Oxford University Press.

Quartermaine, P. and Peter, B. 2006. *Cruise: identity, design and culture.* New York: Rizzoli.

Stiegler, B. 2010. *For a new critique of political economy*. Cambridge: Polity Press.

Strathern, M., ed. 2000. *Audit cultures: anthropological studies in accountability, ethics and the academy*. London: Routledge.

Suchman, L. 1987. *Plans and situated actions: the problem of human-machine communication*. Cambridge: Cambridge University Press.

Thrift, N. 2004. Movement-space: the changing domain of thinking resulting from the development of new kinds of spatial awareness. *Economy and Society*, 33: 582–604.

Thrift, N 2008. *Non-representational theory: space, politics, affect*. London: Routledge.

Turkle, S. 1984. *The second self: computers and the human spirit*. London: Granada.

Urry, J. 1990. *The tourist gaze: leisure and travel in contemporary societies*. London: Sage.

Urry, J. 1995. *Consuming places*. London: Routledge.

Väliaho, P. 2014. *Biopolitical screens: Image, power, and the neoliberal brain*. Cambridge, MA: MIT Press.

Winograd, T. and Flores, F. 1986. *Understanding computers and cognition: a new foundation for design*. London: Intellect Books.

2 "Spoiled", "bored", "irritated" and "nervous"

The transformations of a mobile subject in airport design discourse

Anna Nikolaeva

Introduction

5 November 2011, Amsterdam Airport Schiphol: The Amsterdam–New York flight is scheduled to depart in thirty minutes. The security control takes place right in front of the gate, and people are following the familiar routine without questions. Having passed through the body scanner, I am rearranging my belongings that have just reappeared on the security conveyor belt. At the same time, the security officer asks the passenger behind me to open his hand luggage. I hear the officer exclaiming something as if in surprise and cast a glance at the contents of the open bag. A folded parachute. The officer asks why the man has a parachute and the man gives a perfectly logical answer: "To jump off a plane". "Not this one, I hope?", the officer continues. "No", the man smiles, "I am going to Florida, and I'll be jumping from a plane there". The officer laughs and nods the man signalling that the examination is over.

(Author's fieldwork diary 2011)

This small anecdote illustrates what research on airports has begun to address over the past ten years: the unavoidable diversity and unpredictability of people's behaviour and experiences at airports—spaces that have often been discussed by scholars as intensively controlled and surveilled spaces (Adey 2004; Lyon 2008) and have been compared to prisons (Kellerman 2008; Molotch 2012). Earlier discussions of airports in human geography and anthropology emphasised the passivity of the whole experience of air travel and the infantilisation of the public that is supposedly transformed into a homogeneous crowd obeying identical orders (Augé 1995; Harley and Fuller 2004; Rosler 1998). More recent work has started to tease that image apart by signalling the varieties of people's experiences at airports and by providing more reflexive accounts of what kind of behaviours and interactions airports might encourage due to security regulations and design arrangements. The focus specifically has been on the interests of airport operators who exercise influence on passenger behaviour through agreement or conflict with other parties (Adey 2007; Lyon 2003, 2008; Salter 2008a, 2008b). In this top-down perspective the compliance of passengers with the airport design and operational set-up has rarely been called into question. Some recent studies support the image of a more active (though often cautious) traveller who is aware of their surround-ings, may choose what to do at an airport, makes use of airport facilities in ways that are not necessarily expected by the airport authorities and who may disrupt the monotonous routine of interrogations at the border control by a joke (Fretigny 2013; Molotch 2012). In the context of studying aeromobility, Budd (2011) has pointed to the importance of understanding air travel experiences as embodied and affective and seeing passengers not "as inert 'pax' but as living, breathing, human subjects" (Budd 2011, 1012). Other scholars have begun to discuss the meanings and experiences of passengering (Adey *et al.* 2012) and have reflected upon the socialities of travelling (Bissell 2010).

While bottom-up studies centred upon the figure of the passenger promise a better understanding of people's experiences of spaces of transit, one question is left open: how are people and their behaviour accounted for in airport design? Are they indeed imagined as passive and compliant, as easy prey of the offer of walk-through stores or as cogs in the airport mobility machinery? The chapter approaches this question through an analysis of the views of the parties involved in airport design on the figure of the passenger, thus bringing together a top-down perspective and a subject that is usually dealt with through a bottom-up approach. The chapter draws upon interviews with airport architects, design-ers and managers who play key roles in design process at Amsterdam Airport Schiphol. They would often directly discuss passengers' needs and behaviour in order to rationalise their design solutions as well as indirectly address the subject in their reflections on other topics. The analysis of interviews is supplemented by the analysis of documentary sources, such as annual reports, press releases and architectural and design reviews.[1]

The approach and methodology are informed by the work of sociologists and geographers who unpack the complexity of interactions behind the production of built environment by professionals (Cuff 1992; Franck and von Sommaruga Howard 2010; Yaneva 2009), shift the focus from the architectural form to the process of co-production of buildings by architects and non-architects alike (Imrie and Street 2009), explore its discursive nature (Markus and Cameron 2002) and study buildings as a spatial machinery that expresses and maintains social order (Dovey 1999). Of specific relevance is Cuff's (1989) exploration of the image that architects have of the people who will be using their buildings. This issue is particularly intriguing in the case of public buildings when architects do not know those people in person. Cuff calls them "phantom actors" who are key imaginary figures in architects' thinking and yet are "difficult to keep in focus because of their kaleidoscopic transformations" (1989, 100).

In what follows, I first briefly introduce Amsterdam Airport Schiphol, the key parties in design process and their priorities. I discuss Schiphol's transformation over the last twenty years into "more than just an airport" (Schiphol Group 2009), a process that one could observe at many other airports. This transformation is key for understanding the motivations and arguments of different key stakehold-ers in the airport design and management process. Next I outline disagreements between parties over the state of mind, behaviour and needs of passengers, linking the argumentation of the parties to their agendas and their ways of measuring their success. The conclusion reflects upon the consequences of such disagreements amongst design professionals and foregrounds the questions that this paradoxical and impossible portrait of a mobile subject evokes.

"Don't worry, be happy": how a passenger became a consumer

Amsterdam Airport Schiphol is Europe's fifth busiest airport by passenger traffic with a total passenger volume of about 55 million in 2014 (Schiphol Group 2014). About 40 per cent of all passengers are transfer passengers for whom the Netherlands is neither a destination, nor a point of origin. Schiphol is famous for its non-aviation-related facilities including, for example, a casino, a museum branch, the Airport Library and Airport Park. The airport has received a number of international awards for its distinctive facilities as well as the experience of transit. In 2013 Schiphol became the first European Airport to make it to the top three airports of the world in a SKYTRAX survey of 12 million passengers (SKYTRAX 2013). In 2015 it was ranked the world's fifth best airport in the category "50 million pax+ per year", fourth in the category "Best Airport Leisure Amenities" and seventh in the category "Best Airport Shopping" (SKYTRAX 2015). The portfolio of non-aviation-related facilities of Schiphol may be impressive but at the time of writing in 2016, their offer hardly looks exceptional. Airports worldwide feature art exhibitions, cinemas, fitness centres, hotels, spas and so forth. Schiphol Group, the company that runs Schiphol, however, claims that they are pioneers in creating and branding the airport as a multi-functional city-like environment—an *AirportCity* (e.g. Schiphol Group 2010).

Key roles in design process at Schiphol are played by the Department of Passenger Services, the Department of Commercial Services and Media, security professionals, managers or units at Schiphol who perform design project management and review, architectural supervisors, architects, interior designers and way-finding designers. These parties are the carriers of distinctive sets of ideas about what kind of experience people should have at the airport and how architecture and interior design should ensure that (see also Nikolaeva 2012). Depending on the project other stakeholders may be involved, such as airlines or shop owners.

While most of the parties involved in design at Schiphol recognise the value of efficient passenger operations, the *Passenger Services* department directly and consistently bears responsibility for facilitating people's movement from point A to point B at the airport both in day-to-day management and in planning for the future. While the former means controlling the quality of operations 24/7 (floor managers work at this level), the latter entails being involved in (re)envisioning parts of the terminal through design year by year. This "operational"[2] or "passenger processes" vision—as it is labelled by other parties—is a result of decades' long accumulation of research and expertise and ensuing values and rules within the airport design industry itself, whereas, for instance, commercial and security expertise have a broader range of sources of origin.

The principal goal of the *Commercial Services and Media* department is to earn money for the airport company, as taking care of passenger operations alone has never been a profitable business. Further, because Schiphol is considered to possess a monopoly as an international airport in the Netherlands, according to Dutch law its aviation-related income is subject to regulation which makes it impossible to increase the charges for the airlines above certain limits (ICAO 2013). The non-aeronautical income is not subject to regulation since Schiphol holds no monopoly on retail or food and beverage business. After the terminal facilities were considerably extended in the first half of the nineties, space became available to develop a non-aviation portfolio and, thus, raise profits. Moving people "from A to B" still was the "core business" but a vision of the airport as a "commercial environment" began to evolve—a space that brings money to the operator and at the same time provides entertainment. The *AirportCity* concept came into use around the same time. According to a few former commercial developers at Schiphol, this concept largely boils down to the idea *"Don't Worry, Be Happy"* in which the vision of Passenger Services provides the first part and the commercial developers have to take care of the second. The idea that "efficient processes" were not only not bringing *money* but also not delivering passengers an *experience* for which they would want to return gradually was making its way into design. From the middle of the nineties the process of developing and conceptualising the "added value" of the airport gained full steam. The massive expansion of facilities, the growth of passenger traffic and the new business goals all contributed to this shift. According to a number of interviewees, after expansion works in the early nineties, a new kind of thinking about the airport was possible.[3] Commercial opportunities became more apparent and the authorities were eager to seize them. Since then the Commercial Services and Media department became a more important player in the design process and began to contest ideas and decisions of the Passenger Services department.[4]

Compared with the Commercial Services and Media and the Passenger Services departments, security professionals play a different role in the design process. Seen through the eyes of managers and architects, safety and security maintenance is not a function of the airport but a context in which the main functions are performed. The security apparatus at Schiphol is extraordinary complex and comprises public and private parties that operate within a context of national and international agreements and policies (Amsterdam Airport Schiphol 2013a, 2013b; Schouten 2014). An analysis of the design process shows that in design projects not directly related to border and security control the role of security professionals is limited: they are more involved at the implementation stage rather than during the early stages of envisioning how the space will look and work. Security professionals may occasionally be involved in negotiations with other parties of the kind with which Passenger Operations and Commercial Services often find themselves busy (e.g. about the number of square metres available for a particular facility). The importance of security for design process in projects not directly related to border control and security check is in the impact of these procedures on the traveller: the "don't worry" part of the slogan is to a great degree about rectifying the negative impact of security procedures.

The last party to be introduced is a very important and diverse one: architects and designers. While a variety of architects and designers have worked for Schiphol, a few stand out due to their long-term involvement and broader powers.[5] The key figures are an architectural supervisor, who advises on the urban, architectural and design development of Schiphol, and the chief terminal architect. The former position is temporary and has been held by a number of people during the period covered in this chapter. The latter, the position of the chief architect of the terminal, belongs to Jan Benthem of *Benthem Crouwel Architects,* who has been continuously involved in designing, planning and advising at Schiphol since the end of the eighties. At the moment of writing there is no overseeing designer. Previously this position was held by Nel Verschuuren, who worked at Schiphol since the sixties and left the post in 2005. Since 1967 the ideas of architects and interior designers have had a strong influence on principles which are still foundational for the vision of Passenger Services and to a certain degree are accepted by other parties. Yet, as the airport authorities came to focus more on raising non-aviation revenues, other parties, primarily Commercial Services began to challenge their role.[6]

Finally, an important player is the *Market Research and Intelligence* department. Although it does not participate in the design process at all, its reports on passengers' needs, preferences and wishes are supplied to every party in the design process and thus influence the course of decision-making process. According to the head of the department they are busy making sense of the transformation of a transportation facility into "more than just an airport" whereby a previously frightened and excited *passenger* turns into a *consumer*:

> Nowadays consumers are spoiled and bored. Two, three times a year on average we get on the plane, and as experienced consumers we wish to be treated well. But we have rushed to the airport, have been held up in traffic, have been nervous with all the hassle around the check-in, irritated by lethargy at passport and security control and stressed by the long wait at the counter of the tax free shopping.
>
> How does Schiphol Airport still keep the appeal of its glorified "bus station" to that blasé visitor? The new Schiphol experience is "Do not worry, be happy!" And we are still investigating it.
>
> (Martens 2002, translated from Dutch by the author)

According to Martens (2002), the studies supplied by the department do not provide definitive answers: different parties would use them differently, in accordance with their goals and principles. Believing in the "Don't Worry, Be Happy" idea unites as much it divides, since stakeholders pursue different goals and have different visions of people's needs, capacities and desires. Where worrying stops and being happy begins is a deeply contested subject. As the head of the Market Intelligence department suggests, the mobile subject keeps changing and, thus demands special attention. The next section reveals that the image of mobile subject does not only change through time but also at any given moment it incorporates contradictions and even acquires moral duties as airport design is being contested by disagreeing parties.

The mobile subject: stressed and confused or calm and confident?

The "Don't Worry, Be Happy" principle was first used by Schiphol's interior designers in the 1960s after they had observed passenger behaviour. It was not formulated as a branding slogan, but the idea clearly underpinned much of the thinking in the design process. This can be deduced both from interviews with the current architects who studied the philosophy behind earlier designs and from, for example, a detailed description of the terminal opened in 1967 (that still forms part of the current terminal) written by a British journalist and entitled "Schiphol Puts Passengers First" (Hughes-Stanton 1968). The principle was based on a belief that calming down stressed passengers by ensuring easy way-finding was the priority, and, therefore, in key spaces related to passenger operations no commercial activities or signage were allowed. Shopping, eating out, advertisements and suchlike were deemed of secondary priority which only deserve a passenger's attention once basic procedures related to flying have been completed:

> [The terminal] is designed with the overall aim that passengers may move easily and logically to where they need to go. The main departure hall, for instance, has no kiosks for the sale of magazines or confectionery, no restaurants, and no advertisements to compete with directional signs. It has no special attractions for visitors, and is therefore not unnecessarily crowded . . . If passengers have time to spare, they can go to the top floor restaurants or to the sales kiosks in the arrival area below - here the austerity of the departure level is replaced by an exhilarating, even Pop-art explosion of advertisements.
>
> (Hughes-Stanton 1968, 48, 50)

More than forty years later an architect from Benthem Crouwel working on the lounge redevelopment project reinforced this belief: you need "to get passengers relaxed and calm and then they will go shopping. Otherwise, they won't. They are just running around stressed: 'Where do I have to go?' They don't even see the shops!"

While Commercial Services and Media agree that this type of thinking is largely correct—a lost and stressed passenger would hardly pay attention to shops or entertainment—they often contest the solutions presented by architects as logically ensuing from this belief. Both parties use assumptions about people's motivation and behaviour to support argumentation.

First, there is disagreement on the degree of stress passengers experience and about their capacity to handle it. Passenger Services and architects try to protect passengers from 'distractions,' arguing that big bright advertisements, a lot of visual diversity and stimulation prevent people from calming down and concentrating on finding their way. Commercial managers agree that the airport should not be a "complete mess" but doubt whether someone would miss a flight if they introduced more visual diversity in the terminal by, for instance creating "more space for retailers to express their shop". A manager who worked on the "border" between the departments believes that stress and its impact on people's decision-making capabilities is overrated:

> I think people have changed …. If you look back on old commercials, on TV… from the seventies …. One single story-line and it ends with a clear conclusion that this soap is the best soap. And now we are used to having different story-lines mixed up together in a real staccato and fast way … and two or three of these stories count ten seconds. Original commercials were … [up] to 60 seconds. So we are used to get much more information per second than 30 years ago. …The passengers … or any people are capable of picking out those elements of information that they need.

Proceeding from these different visions, the parties pursue different strategies in the struggle for orchestrating people's attention. Passenger Services constantly work on removing all the information they consider "extra", while Commercial Services and Media invent new ways of exposing the passenger to *their* information: leaflets, special promotions with staff, visual projections on the floor, apps, Facebook page, Schiphol TV: that is, *expanding* in physical and virtual space.

The second significant difference in thinking about the behaviour of people at airports is the perspective on their needs. In the view of the architectural supervisor between 1996 and 2008, Hubert-Jan Henket, "you don't want to be distracted" by shops or advertising. He believes that people want to go to the gate, and if they long for distractions—they should find them in Disneyland, not at an airport, so, in effect, they *should not* want to shop. Jan Benthem, the terminal chief architect, does not take an issue with the very existence of shops but searches for ways to give people choice, relying on the idea that if people can see options they will choose doing what they prefer most. His ideal passengers seem to be more inclined to shop but are not too easily "seduced" by different "tricks" of the commercial managers: they do not buy things "accidentally" just because they are exposed to an offer: "I think it doesn't work because you have to be in a state of mind to go shopping. … You can't force people to go shopping. … We disagree with the airport authorities sometimes about it". Therefore, according to Benthem, the "walk-through" stores which are now so common at airports are confusing. They deprive people of the chance to choose: immediately following security control, passengers find themselves in a store through which they should pass before reaching the gate. For instance, airports in Vienna, Manchester, Saint-Petersburg, Budapest and Copenhagen have such a set-up.

Commercial Services and Media acknowledge that the "first need" of the passenger is going through flying-related procedures. Shopping and having fun is their "second need" in which some have more interest: passengers even may be eager to search for, specific brands such as Victoria's Secret or Starbucks. Others have to be given opportunity—and then they might make use of it, according to commercial developers. The goal is to create plenty of such opportunities and make them attractive:

> [Shops] are there, because [passengers] want it. Of course, I will do my utmost to make it more interesting to go in, that's the exactly my job. I can never make someone go in, if they really don't want to. It's a need of the passenger, and they expect it to be there!

The ways of measuring success and the attitudes about people's behaviour are closely intertwined. As the goal of Passenger Services is to reach more than 90 per cent of recognition of signage, they imagine a passenger who is easily distracted or cannot manage too much information, thus—commercial developers would argue—possibly, overprotecting others. Commercial Services and Media, on the other hand, wish to activate the supposed 'second need' in the maximum number of people. Commercial Services and Media cannot boast about 'reaching' their goals, but rather continually experiment to expand the range of means that might make people consider going shopping.

Thus, all parties strive to guide the passenger either way: to the gate or to the shop, interpreting human behaviour in accordance with their own vision and wishing to encourage particular behaviours for profit or being disappointed that people do not always behave as they 'should'. Airport design is a contested field rather than an implementation of a single vision, and, hence, the image of the passenger also remains highly contradictory—despite attempts to 'fix' passenger identity through market research.

Conclusions

Mobile subjects in airport design process become chess figures in multiple games for space, for profit and for efficiency, undergoing transformations and eluding definitions. It is this unpredictability and mysteriousness of these "phantom actors" that both puzzles airport designers and managers and grants them the room for manoeuvre. Most interviewed professionals believe in the "Don't Worry, Be Happy" principle: first the potential reasons for the passengers to be worried have to be eliminated, and only then the leisure amenities can spark their interest. Yet this slogan is as much a "formula" as an equation with two unknowns: ideas about the passengers' resilience to stress and ability to plan the route from A to B as well as their interest in shopping or entertainment are a subject of ongoing contestation.

According to Markus and Cameron, "there are no 'innocent', power-free spaces" (2002, 69). Architectural briefs and any communication pertaining to the process of designing a building "contain some prescriptive element, and … all have an ideological dimension" (2002, 41). Each image of the passenger mobilised in the process of contestation of airport design has its implicit or explicit ideological foundations and wide-reaching practical implications. Should an airport remain as much as possible a travel-related space without extras because passengers are too stressed or even incompetent to handle extra choices or possible temptations? Should passengers be protected from extra information that could be distracting or they can be trusted in their choices of the route through the airport, speed of movement and activities on the way from A to B? Whilst appearing practical in nature, such questions have clear moral undertones.

Airports are spaces of temporary dwelling, but they are tricky to leave once inside, and if one wants to reach their destination the "house rules" should be followed. But does one become a non-smoker just because it is cheerfully announced throughout the terminal that it is a non-smoking airport? The choice to get rid of smoking rooms may be driven by an operational objective (to free more flow space for example) or a commercial project (adding extra shops), but it is also unavoidably a moral decision whereby airport authorities impose a rule related to healthcare rather than to travel on millions of foreign citizens. Obviously, cigarettes and alcohol would still be sold in impressive quantities in duty free stores.

This chapter offers a new perspective into existing research on the governance of mobile subjects and the ambiguity of the figure of the passenger (Adey 2008; Adey *et al.* 2012). The investigation of ideas held by architects and designers about passengers reveals that for airport-makers, passengers can be *simultaneously* "pax" (Budd 2011) whose routes are prescribed, stressed individuals who are always at risk of becoming too confused to be in control as well as subjects choosing their activities as they please and navigating the airport with confidence. Each of these dimensions of a mobile subject is mobilised to justify designs serving particular goals and achieve versions of success that different groups of "airport-makers" strive towards.

Notes

1 In total, forty interviews were conducted between 2010 and 2011. The brief historical narrative in this chapter is based on the extensive analysis of annual reports and other documentation, conducted within the framework of a bigger research project on designing airports as multifunctional public spaces and on Schiphol as an "AirportCity" in particular. For more details on designing the airport as a space for mobility and temporary dwelling see Nikolaeva (2012), for a detailed discussion of Schiphol as an AirportCity see Bosma and Nikolaeva (2013a, 2013b).

2 Hereafter the source of direct quotations is the interview material. Interviewees are anonymised except easily identifiable architects and architectural supervisors who have given their permission for their name to appear in publications.

3 For a detailed discussion of the expansion see Berkers and Burgers (2013) and Kloos and de Maar (1996).

4 For an analysis of "struggles" and "negotiations" between different parties at Schiphol see Nikolaeva (2012).

5 Architecture and design at Schiphol are the subject of a number of publications written both by researchers and professionals who have worked for Schiphol. For historical accounts of architecture and design and Schiphol see Bosma (2013), Kloos and de Maar (1996) and van Beusekom and Huygen (2005).

6 For a discussion of disagreements on commercial developments see Henket (2013), Nikolaeva (2012), Verschuuren (2005).

References

Adey, P. 2004. Secured and sorted mobilities: examples from the airport. *Surveillance and Society* 1: 500–519.

Adey, P. 2007. 'May I have your attention': airport geographies of spectatorship, position and (im)mobility. *Environment and Planning D: Society and Space* 25: 515–536.

Adey, P. 2008. Airports, mobility and the calculative architecture of affective control. *Geoforum*, 39: 438–451.

Adey, P., Bissell, D., McCormack, D. and Merriman, P. 2012. Profiling the passenger: mobilities, identities, embodiments. *Cultural Geographies*, 19: 169–193.

Amsterdam Airport Schiphol. 2013a. *Private parties.* Available at http://extra.aviationonline. schiphol.nl/Home/SecurityAndSchipholPass/PrivateParties.Htm, accessed 12 July 2013.

Amsterdam Airport Schiphol. 2013b. *Public organisations—the role of government and legislation.* Available at http://extra.aviationonline.schiphol.nl/Home/SecurityAnd SchipholPass/PublicOrganisations.htm, accessed 12/7/2013.

Augé, M. 1995. *Non-places: introduction to an anthropology of supermodernity.* London: Verso.

Berkers, M., and Burgers, I. 2013. Structuring masses: architecture in a race against time. In *Megastructure Schiphol: design in spectacular simplicity*, Ed. K. Bosma. Rotterdam: nai010 Publishers, pp. 252–269.

Bissell, D. 2010. Passenger mobilities: affective atmospheres and the sociality of public transport. *Environment and Planning D: Society and Space*, 28: 270–289.

Bosma, K. (ed.) 2013. *Megastructure Schiphol: design in spectacular simplicity.* Rotterdam: nai010 Publishers.

Bosma, K. and Nikolaeva, A. 2013a. Farewell to spectacular simplicity? in *Megastructure Schiphol: design in spectacular simplicity*, Ed. K. Bosma. Rotterdam: nai010 Publishers, pp. 298–306.

Bosma, K and Nikolaeva, A. 2013b. The airport: prototype of the global city? In *Megastructure Schiphol: design in spectacular simplicity.* Ed. K. Bosma. Rotterdam: nai010 Publishers, pp. 198–217.

Budd, L. 2011. On being aeromobile: airline passengers and the affective experiences of flight. *Journal of Transport Geography*, 19: 1010–1016.

Cuff, D. 1989. Through the looking glass: seven New York architects and their people. In *Architects' people.* Ed. R. Ellis and D. Cuff. New York: Oxford University Press, pp. 64–102.

Cuff, D. 1992. *Architecture: the story of practice.* Cambridge, MA: MIT Press.

Dovey, K. 1999. *Framing places: mediating power in built form.* London: Routledge.

Franck, K.A. and von Sommaruga Howard, T. 2010. *Design through dialogue: a guide for clients and architects.* Chichester: Wiley.

Fretigny, J-B. 2013. *Les mobilités a` l'épreuve des aéroports: des espaces publics aux territorialités en réseau. Les cas de Paris Roissy-Charles-De-Gaulle, Amsterdam Schiphol, Francfort-sur-le-Main et Dubai International.* Paris: Université Panthéon-Sorbonne.

Harley, R. and Fuller, G. 2004. *Aviopolis: a book about airports.* London: Black Dog Publishing.

Henket, H-J. 2013. The only thing that is permanent at Schiphol is change. In *Megastructure Schiphol: design in spectacular simplicity*, Ed. K. Bosma. Rotterdam: nai010 Publishers, pp. 78–79.

Hughes-Stanton, C. 1968. Schiphol puts passengers first. *Design* 240: 48–55.

ICAO. 2013. *Case study on commercialization, privatization and economic oversight of airports and air navigation services providers: Netherlands.* Available at http://www. icao.int/sustainability/CaseStudies/Netherlands.pdf, accessed 15/7/2013.

Imrie, R. and Street, E. 2009. Regulating design: the practices of architecture, governance and control. *Urban Studies*, 46: 2507–2518.

Kellerman, A. 2008. International airports: passengers in an environment of 'authorities'. *Mobilities,* 3: 161–178.

Kloos, M. and de Maar, B. 1996. *Schiphol architecture: innovative airport design.* Amsterdam: Architectura & Natura.

Lyon, D. 2003. Airports as data filters: converging surveillance systems after September 11th. *Journal of Information, Communication and Ethics in Society* 1: 13–20.

Lyon, D. 2008. Filtering flows, friends and foes: global surveillance. In *Politics at the airport.* Ed. M.B. Salter. Minneapolis: University of Minnesota Press, pp. 29–49.

Markus, T. A. and Cameron, D. 2002. *The words between the spaces: buildings and language.* London: Routledge.

Martens, H. 2002. 'Het Museum als Luchthaven?' Available at http://www.gerbenmarcel jacobs.com/MasterclassMarktonderzoek.html, accessed 21/6/2013.

Molotch, H. L. 2012. *Against security: how we go wrong at airports, subways, and other sites of ambiguous danger.* Princeton, NJ: Princeton University Press.

Nikolaeva, A. 2012. Designing public space for mobility: contestation, negotiation and experiment at Amsterdam Airport Schiphol. *Tijdschrift voor economische en sociale geografie* 103: 542–554.

Rosler, M. 1998. *In the place of the public.* Berlin: Cantz.

Salter, M. B. 2008a. Introduction: airport assemblage. In *Politics at the airport.* Ed. M. B. Salter. Minneapolis: University of Minnesota Press, pp. ix–xix.

Salter, M, B. 2008b. The global airport: managing space, speed and security. In *Politics at the airport,* Ed. M.B. Salter. Minneapolis: University of Minnesota Press, pp. 1–28.

Schiphol Group. 2009. *From airfield to AirportCity.* Available at: http://www.schiphol.nl/ SchipholGroup/NewsMedia/SchipholFactSheets.htm, accessed 18/8/2013.

Schiphol Group. 2010. *Innovative from the start.* Available at http://www.schiphol.nl/ SchipholGroup/NewsMedia/SchipholFactSheets.htm, accessed 1/8/2010.

Schiphol Group. 2014. *Annual report.* Available at: http://www.annualreportschiphol. com/about-us, accessed 8/4/2015.

Schouten, P. 2014. Security as controversy—re-assembling the security apparatus at Amsterdam airport. *Security Dialogue*, 45: 23–42.

SKYTRAX. 2013. *Airport award winners.* Available at http://www.worldairportawards. com/Awards_2013/category.htm#terminal, accessed 12/7/2013.

SKYTRAX. 2015. *Airport award winners.* Available at: http://www.worldairportawards. com/Awards/airport_award_winners_2015.html accessed 7/4/2015.

van Beusekom, F. and Huygen, F. (eds). 2005. *Flow. Het Schiphol van Nel Verschuuren 1968–2005.* Amsterdam: Schiphol Group.

Verschuuren, N. 2005. Interview by F. Beusekom. In *Flow. Het Schiphol van Nel Verschuuren 1968–2005.* Ed. F. van Beusekom and F. Huygen. Amsterdam: Schiphol Group, pp. 43–48.

Yaneva, A. 2009. *Made by the office for metropolitan architecture: an ethnography of design.* Rotterdam: 010 Uitgeverij.

3 Legible London

Mobilising the pedestrian

Spencer Clark, Philip Pinch and
Suzanne Reimer

Introduction

This chapter examines the design of the Legible London pedestrian wayfinding system. Overseen by Transport for London (TfL), this innovative scheme for enabling walking has developed from an early prototype study in 2007 to become a key part of transport policy in the UK's capital city (AIG 2006, 2007; Arquati 2008; TfL 2014). An integrated combination of signs, pedestrian focused mapping and other directional information, Legible London has two complementary aims; to help people plan journeys on foot; and to give people the confidence to walk and explore. The scheme consists of a city-wide, consistent, pedestrian navigation system encompassing on-street wayfinding elements supported by identical information in public transport nodes (e.g. tube stations and bus stops) and paper-based products as well as ongoing development of the provision of digital mapping information. The current on-street system comprises a mixture of information boards, known as 'liths' that come in a standardised range of sizes (see Figure 3.1), directional fingerposts, wall-mounted signs and a range of supporting printed maps located in bus stops and inside tube stations. Information on all liths is presented in a hierarchical fashion: a top yellow beacon locates the sign in busy urban environments, directional information is given to nearby points of interest (replicating traditional finger signs), whilst differently scaled 'planner' and 'finder' maps locate the lith within 15-minute and 5-minute walk scales respectively. A street index also is provided.

The Legible London scheme has its origins in research conducted by the London School of Economics on behalf of the consultancy Applied Information Group (AIG),[1] which sought to understand existing wayfinding provision in Central London (Ichioka *et al.*, 2005). Using this evidence base, AIG was commissioned by the Central London Partnership (a group of central London boroughs) to research potential barriers perceived by pedestrians, "to propose ways of dealing with them" and to understand how levels of walking in the capital might be increased (AIG 2006, 2007; Arquati 2008, 2). One of the most significant barriers was found to be wayfinding: a range of surveys reported that between 50 per cent and 60 per cent of Londoners would walk more if they had better wayfinding information (Clark 2008; Arquati 2008). Prior to the Legible London scheme,

Figure 3.1 Legible London 'monolith', Oxford Street.

information systems across London had been isolated, incomplete, and lacked consistency, which was seen to discourage people from undertaking walking journeys.

The central aim of this chapter is to critically review the shape and nature of the Legible London scheme, highlighting the ways in which design decisions shape how and where people move around the city. Our discussion draws upon reports assessing the first Legible London prototypes trialled in Bond Street, central London (AIG 2007; TRL 2008; Colin Buchanan 2008); upon subsequent evaluations of the scheme (TfL 2014); and upon primary research conducted as part of Clark's (2008) MA dissertation at London South Bank University. We begin by outlining the factors which led to the development of the scheme. The main body of the chapter considers the ways in which the Legible London initiative has sought to 'order' wayfinding in the capital, redefining how, where and who moves by suggesting destinations to pedestrians; installing information signs which are readable at specific vantages and locating this information at particular points in urban space. In doing so we consider the ways in which forms of physical mobility such as walking become meaningful (Cresswell 2006). A final section reflects upon the mobilisation of design by considering the travels of the Legible London scheme itself, as it has moved to become a template in a range of other cities across the globe.

Origins of the scheme

A key driver for the scheme was TfL's interest in alleviating a growing public transport burden. There is at one level a strong irony in the idea of a local government transport agency directly seeking to encourage people to opt out of its services! However, significant increases in London's population have led to overloading of the transport network, leading to disrupted and less pleasant journeys for commuters. The Greater London Authority (GLA 2016) reports that London's population has increased by around 113,000 people per year over the past five years, and is estimated to surpass 9 million in 2018 and 10 million by 2034, leading to continuing pressure on public transport infrastructure. Thus the Legible London initiative can be seen as a fascinating mobilisation of the pedestrian—that is, TfL has seen a need for the pedestrian to take on a public transport role.[2] Walking has become conceptualised as a service in which TfL is centrally involved (TfL 2004). This vision more recently has been extended to cycling through bike-share and 'cycle superhighways' schemes and related infrastructure investments (GLA 2013).

Initial analysis by the consultancy AIG sought to ascertain factors that might prevent people from walking in London. Surveys (Clark 2008; AIG 2007) indicated that safety concerns (including the perceived danger of traffic congestion) typically presented barriers to walking. Wayfinding difficulties also were key, as we have indicated. More specifically, a consideration for both residents and tourists was the fact that the iconic London tube map was so frequently used as a wayfinder. In a survey of pedestrians exiting Leicester Square tube station in March 2005, AIG found that 45 per cent used the tube map to plan their journey in advance (Ichioka *et al.* 2005; see also Bozatli *et al.*'s 2004 New York research). Whilst the tube map is highly successful in assisting users to find a station and locality, it cannot guide them beyond the station itself for the potential onward walking leg of their trip. Further, the iconic London tube map of course is neither to scale nor geographically correct and as a result often encourages very short public transport journeys. For example, although many central London stops are just minutes apart, the tube map gives no sense of this. Thus whilst the tube map occupies a powerful position in people's decision to wayfind with available tools, they are typically unaware of the tube map's unsuitability for pedestrian navigation. The lack of suitable pedestrian focused alternatives—and the fact that existing systems were not fulfilling their intended purpose—were crucial factors in the design and implementation of Legible London.

Theorising wayfinding: order and pedestrian mobilities

One of the most interesting elements of the Legible London scheme is the extent to which it explicitly seeks to order pedestrian experiences and behaviour. This impetus for order derives from the ways in which wayfinding is understood and in this respect the scheme overtly draws upon urban planner Kevin Lynch's (1960) understanding of mental mapping (Clark 2008, 10). Lynch (1960) classified the content of city images into five formal types of 'image elements' (paths, landmarks, edges, nodes and districts), which he suggested were used by people to order their understanding of urban built environments and to construct mental pictures to cue their navigational strategies. Edges denote boundaries between districts, which are areas that share common characteristics. Nodes are places of convergence and landmarks are distinctive objects. Lynch's (1960) emphasis on the importance of clear "legibility" in the urban environment is unmistakeably evident in the naming of the London scheme.

Many discussions of wayfinding suggest that environment familiarity plays a key role. Golledge (1992) notes that people appear to prefer following repetitive routes that match their mental maps rather than exploring alternatives; whilst Xia *et al.* (2008) note that for unfamiliar environments, people use landmarks with support from signage, of which the latter is considered especially important. Landmarks may not always be used in the same way: Passini (1981) suggests that some individuals rely on signage to a greater extent and wayfind in a linear fashion, whilst others navigate spatially making greater use of the general environment. Users also are seen to be more reliant upon signage in environments that are unfamiliar (Passini 1981); and there has been some discussion about gender differences in wayfinding strategies (Schmitz 1999; Xia *et al.* 2008). General assumptions about the central importance of environmental landmarks can be seen to have fed directly into Legible London's development of pedestrian wayfinding systems that utilise landmark features.

Predictability and the importance of 'not getting lost' can be seen as key elements of the Legible London strategy, again drawing upon central aspects of Lynch's (1960) work (see also Passini 1996). Being lost is seen to be a frightening experience:

> [L]et the mishap of disorientation once occur, and the sense of anxiety and even terror that accompanies it reveals to us how closely it is linked to our sense of balance and well-being. The very word "lost" in our language means much more than simple geographical uncertainty, it carries overtones of disaster.
>
> (Lynch 1960, 4)

If being lost or simply the fear of being lost dissuades people from walking, it follows that providing improved wayfinding through pedestrian signage systems should encourage people to walk more often. AIG clearly reiterated such a view in their 2008 report:

> In London, the problem of disorientation can be acute. People find many areas hard to understand and this induces considerable stress. The realisation of being 'lost' can be a negative feeling, bring on panic and a sense of impending disaster.
>
> (2008, 13)

Some assumptions about wayfinding 'on foot' are seen to parallel wayfinding whilst driving. Burns (1998) stresses the importance of wayfinding in relation to the main reason for driving, which is to travel and reach a destination safely, conveniently and without assistance from outside sources wherever possible. As soon as a driver gets lost, takes a wrong route or needs to seek assistance, the functionality and mobility of driving decreases. Wayfinding can thus be considered a crucial component of successful and satisfactory driving experiences (Clark 2008, 15–16). The importance of hierarchical wayfinding (Passini 1981) to driving has been a key feature of signage design for the British motorway network, for example.

Despite such apparently dominant assumptions about the need for legibility, predictability, and imageability in the urban environment, it is important to emphasise that such an ordered and indeed more 'rational' approach is not the only way in which wayfinding has been considered. Indeed, developing parallel to Lynch (1960) was Situationist International's experiments with psychogeographic explorations of city environments, which through the concept of the *dérive* advocated calculated attempts to disrupt the psychologies of routinised patterns of urban movement and encounter (Bonnett 1989, Pinder 2004). The disparity between more and less instrumental and rational modes of movement highlights that meanings of mobility are not fixed, but rather take shape within broader political-economic contexts (Cresswell, 2006; Spinney 2016).

Features of the Legible London scheme

The chapter now turns to examine how the Legible London scheme seeks to achieve its key objectives. From the 2007 pilot project onwards, AIG and TfL agreed that the most appropriate method of providing information and thus increase walking in London would be to design a "master" structure around which all system elements would attach. Known as the Living Map, a master structure allows all online maps, printed maps, on-street systems, and at-exit tube maps to present information from a single source for consistency. These elements combine to provide a comprehensive pan-London wayfinding system for pedestrians.

The on-street elements of the Legible London scheme—lith information boards, wall signs as well as finger posts—are in part determined by the physical restrictions of London's built form, with physical space constraints a key consideration. On-street maps must be a certain size and scale to be useful; but at the same time narrow pavements and organic street patterns can constrain the placement of signage. Map keys, text font size and overall clarity must allow use by all: the scheme must meet Disability Discrimination Act (DDA) requirements, for example.[3]

In making decisions about what types of information were to be provided on system elements, AIG initially determined four main questions must be answered:

1   Where am I?
2   Where is it?
3   How do I get there?
4   What else is around here?

Information present on the system elements (maps, text, images, key, etc.) is a direct answer to each of these four questions.

AIG worked with experts in architecture, graphics, built environment and urban design to develop the first prototypes; and user feedback played an important role in system development. Within individual maps, naming conventions and classifications built upon existing districts and villages of London, in part because village names have been long used by bus companies to identify the start and end points of routes. Further, area naming in itself was seen as important:

> Legible London can influence the behaviour of Londoners and visitors, simply by creating awareness that London has a number of named areas and that all you need to find a place is to know what area it is in.
>
> (AIG 2008, 16)

On-street map display is via what is known as 'heads-up mapping', as this was believed to be most easily understood and absorbed by pedestrians and thereby promote "intuitive understanding" (AIG 2007, 48). Heads-up mapping removes the mental transformation (or map turning) needed for orientation that is usually encountered when using traditional north-based (compass) mapping when the user is not actually facing north. Interestingly, it was decided that Legible London maps in tube stations would use north-based maps because such locations have no direct physical connection with the street environment.

AIG believed that Legible London needed to provide consistency and familiarity of information and design, otherwise it would risk repeating mistakes made by (earlier) disparate individual London borough council and Business Improvement District systems, which lacked information cohesiveness and consistency. A range of key features were built in to achieve such consistency: progressive disclosure; reliability and predictability; avoiding 'navigational waste'; visibility and legibility.

Progressive disclosure

Progressive disclosure as a process involves supplying just enough information at each decision point to navigate successfully, without overburdening the user. AIG (2006) investigated existing wayfinding best practice examples from a range of global locations and cited the modern British road and motorway signage system, which uses progressive disclosure of information, as an ideal basis for providing pedestrian information at the right time and place as required. AIG believed that the level of clarity achieved by the UK road sign system was sorely missing from pedestrian wayfinding systems, illustrated by the 32 different pedestrian wayfinding systems in place across Central London in the early millennium (AIG 2006; Clark 2008, 17).

Reliability and predictability

Initial reports stressed the need for reliability and predictability in the Legible London scheme:

Availability of information is crucial to wayfinding decision making .... At a certain point along a route, no information (or only contradictory information) may be available. In this situation you have no other options than to resort to trial and error, making decisions by chance, or perhaps, on instinct.

(Arthur and Passini 1992, cited in AIG 2008, 19)

One of the strengths of walking as a mode is that, once a route is known, it rarely varies in length of journey. However, in contrast to other modes, where an unknown journey can be investigated through timetables (AIG 2006), unknown walking journeys are seen as unreliable in that a timetable is not present. This situation was seen to arise directly from a lack of sufficient pedestrian information to give accurate indications of time to walk. Linked to this, such trips are not reliable in terms of support once walking on-street. Thus AIG concluded that journey time reliability and associated information would encourage more walking trips (Clark 2008, 18).

Navigational waste

The avoidance of 'navigational waste' (Burns 1998) also was seen to be important in the development of the Legible London scheme. This refers to the excess journey time above an optimal route which is lost to wayfinding errors and incorrect decision making. Avoiding navigational waste was deemed particularly important if the scheme was to be successful in encouraging both tourists and commuters to move more quickly and efficiently through urban space. AIG (2008) suggested that time spent navigating, using landmarks to wayfind and rechecking progress, was time lost to observation, thought or other activities, with the consequence that the 'place' is not used, but simply passed through (Clark 2008, 17).

Visibility

A key point of tension in any pedestrian wayfinding scheme is the balance between being able to see signs and wayfinding information, whilst at the same time being able to distinguish elements of the surrounding environment. Indeed one of AIG's criticisms of existing London signage was that chaotic 'street clutter' (e.g. vehicle signs, traffic signals, lamp columns, guardrails and shop awnings) made it difficult for pedestrians to find their way. AIG stressed that:

Information, in particular graphic information, has to be designed for normal environmental perception, which consists of the scanning and glancing process. People tend to ignore information displays that are not designed appropriately, or to walk away from such displays after spending a minimum of time in futile search.

(2008, 28)

Thus new pedestrian wayfinding systems must be distinctive, consistent, offer appropriate and useable information without becoming lost in the general streetscape (Clark 2008, 18–19).

Legibility

The design of individual elements within the Legible London scheme (see also Figure 3.1) has sought to combine elements of colour, clarity and boldness in order to achieve strong legibility. At the top of the wayfinder, a yellow strip acts as a beacon to help pedestrians find the sign, while an address below locates the individual sign and the area in which it placed. In the middle of the wayfinder, directional information is provided to give quick confirmation of nearby attractions or points of interest. In the bottom half, two maps and a street index fill any further information gaps for the reader. Using time as a measure of travel distance, a 15-minute scale planner map provides more general orientation, whilst the 5-minute scale finder map helps the user identify detailed local landmarks. A detailed street index enables pedestrians to identify specific roads and to find their final destination.

The 5-minute distance finder map (see Figure 3.2) explicitly registers the importance of pedestrians' mental maps to the design of the Legible London system. AIG (2008) argued that a pedestrian wayfinding system which used human, as opposed to motor traffic, scales of distance, memory and recognition would engage more effectively with pedestrians and thus be more likely to be used. It was felt that locations within a 5-minute walk distance were liable to seem more

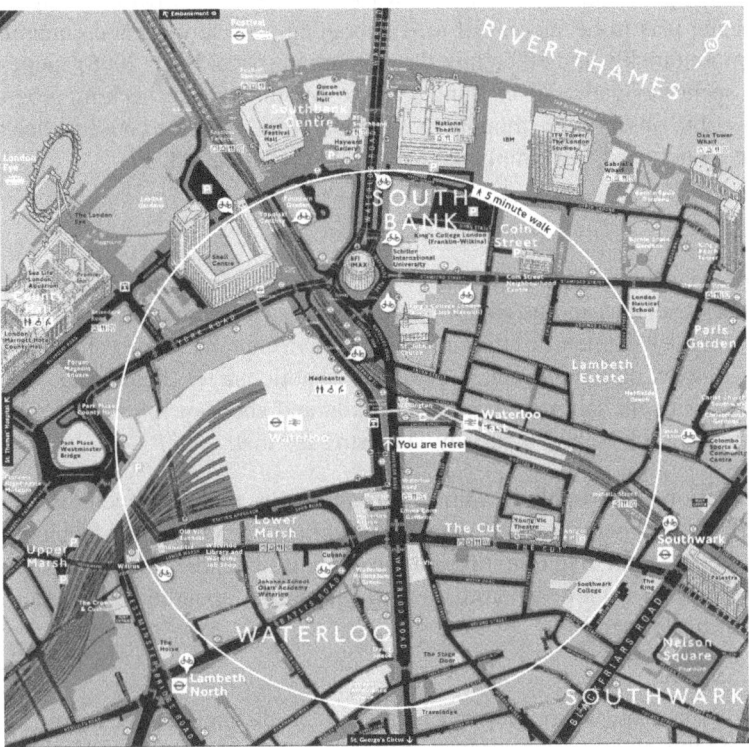

Figure 3.2 South Bank 5-minute 'finder' map.

easily walkable and thus encourage walking. Time rather than distance was used as a measure following an AIG (2006) survey that 73 per cent of respondents could not accurately estimate actual distance and fewer than 60 per cent could correctly identify the direction to another tube station (AIG 2006, cited in Clark 2008, 41).

Although the depiction of journey time rather than distance has a 'user-friendly' intent, it is important to indicate that a generic measure of travel time makes particular assumptions about the types of bodies walking in the city. It is assumed that pedestrians have uniform physical capabilities and comportment; and are not accompanied by young children, for example. In some written documentation (e.g. AIG 2007, 30) a 400 m distance equivalent is given for the '5-minute walk'. But on the ground, the scale of the 5-minute finder map can be seen to give precedence to design simplicity over specified inclusivity.

In other ways, however, the Legible London scheme does seek to be cognisant of bodily differences among users as well as individuals' mental maps. It was emphasised that "a pedestrian wayfinding system should cater for all individuals, whatever their preference for information provision and this will in part be determined by their mental map experiences" (Clark 2008, 11). Individuals acquire and use different sensory clues in the urban environment, which has important implications for the type of wayfinding support that might be provided. Further, as portrayed in Figure 3.3, the design of Legible London signs involved a calculation of Department of Transport standards for *x*-heights (a unit of measurement used to define the point at which text and images in a sign become legible to the viewer) across a range of visual acuities and visual fields.

A study by Department for Transport (DfT) and Walk England (2008) highlighted a number of ordering principles that all walking maps (and by extension, wayfinding systems) should communicate to their intended users; and the Legible London maps draw upon these. The provision of consistency and continuity across borders and boundaries is one element, along with supported encouragement for exploration when needed. Maps should also be able to promote local identity (Clark 2008, 21–22). Figure 3.2 above reveals how a Legible London map registers a local area, in this case the South Bank 'village'; and includes landmarks such as the National Theatre, the Hayward Gallery and Coin Street Neighbourhood Centre. Transport nodes such as Waterloo Station as well as bus

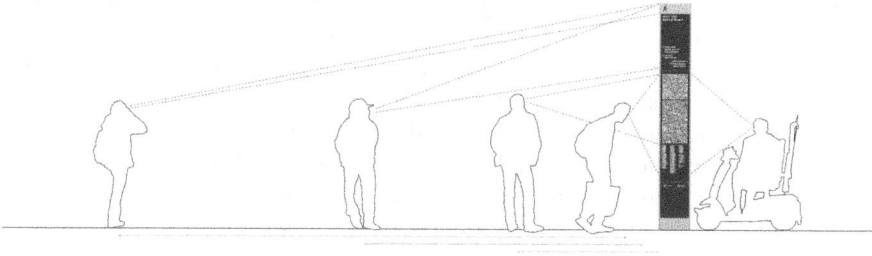

Figure 3.3 Catering for different user tasks, reading strategies and physical distance from the sign.

stops also are prominent. Critical cartography has increasingly drawn attention to the importance of who or what is on the map and the way such features shape how we know and understand environments (Harley 1989; Dodge *et al.* 2011). In the Legible London scheme, landmarks are chosen according to a specific set of criteria, or 'system rules' ranging from memorable, identifiable and internationally recognisable sites, to listed buildings of architectural merit and key 'live assets', such as civic buildings, open spaces, places of worship and monuments. Live assets can also include commercial buildings, such as 'landmark retail' outlets, cinemas and theatres, although unlike marketed systems such as Google maps, organisations cannot pay for recognition.

Implications and conclusions

This chapter has drawn upon the example of the Legible London pedestrian way-finding scheme in order to explore how decisions made by designers can shape how and where people move around the city. By encouraging pedestrians to walk rather than to take public transport, the scheme explicitly has sought to change their mental maps in finding their way through the city. Further, we have seen that wayfinding has been conceptualised as something which is more efficient if it is conducted in an orderly and controlled manner: individual elements of the Legible London scheme seek to provide a unified and coherent design system for pedestrian mobility.

The overall intent of the scheme both drives and brings into conflict different meanings of mobility. At one level, Legible London seeks to enhance urban productivity by enabling labouring commuters to move efficiently through the city. At the same time, the scheme also aims to facilitate more leisurely tourist movement through the city, even though this is a rationalised form of movement. It could be argued that if these two different goals did not need to be reconciled, the version of tourist walking that Legible London promotes might have been influenced instead by a situationist *dérive*, offering some form of 'lucky dip' for those inclined to explore.[4] In broad terms we might further emphasise that the practice of walking itself can provide opportunities for alternative tactics, even as the physical fabric of the street (including Legible London signage) is ordered and stable (Urry 2007, 73; drawing upon de Certeau 1984).

Finally, the scheme is fascinating in the sense that the components, design objectives and underlying principles of Legible London have themselves become mobile: similar schemes have been implemented in UK regional cities such as Southampton and Bristol as well as internationally—for example in Beijing, New York and Sydney, Australia. In part, such design mobility has occurred because the originating 'legible city' consultancy (now called Applied Wayfinding) has been commissioned to undertake further work, ranging from an indoor scheme for the National Gallery in London to a redesign of wayfinding systems across Hudson Yards, a retail and public space in Manhattan, New York City. But also, as the founder and chief designer at Applied Wayfinding has noted:

Legible London-inspired solutions have popped up in Sweden, China, Australia, the U.S., Canada, and Russia, as well as across the UK. They all have similar beacon-style units featuring to-scale maps and heads-up orientation (of course), with integrated directional information. It has become a global standard. A great British export.

(Fendley 2015)

In the emergence of what Fendley (2015) believes to be a "global standard", we can see the ways in which ideas about design travel through different spaces and places. Such mobilities have important effects, shaping how cities are read, understood and moved through. Equally, however, they are likely to be developed, transformed and indeed subject to counter processes of resistance. The recent reawakening of interest in psychogeographic investigations of cities in part has emerged from concern over singular and corporatised representations of urban space, the predictability of urban planning methods and the controlling impulses they embody (Bennett 2011; Garrett 2014). How these counter tendencies mobilise new media and spatial technologies to define and redefine pedestrian experiences will be an important feature of city living in the twenty-first century.

Notes

1 When the Legible London scheme was planned, designed and launched, the consultancy was known as AIG, although more recently its name has changed to Applied Wayfinding: see www.appliedwayfinding.com (accessed 13/06/16). For clarity this chapter uses the original consultancy name (AIG), under which reports and plans for Legible London were first published.
2 There are strong parallels with Spinney's (2016) analysis of the role of the cyclist.
3 For a discussion of the limitations of 'universal design', see Kullman, Chapter 10, this volume.
4 A number of mobile phone apps are now available which seek to help users develop their own *dérive*, although of course there is a certain irony in a reliance upon location fixing technologies whilst also seeking to 'get lost'. On the critical potential for unsettling and un-fixing location amidst the positioning technologies of GPS, see Pinder (2013).

References

Applied Information Group (AIG) 2006. *Legible London: a wayfinding study*. March 2006, AIG and Central London Partnership.
Applied Information Group (AIG) 2007. *Legible London: Yellow Book: A prototype wayfinding system for London*. London: Transport for London/AIG.
Applied Information Group (AIG) 2008. *Legible London: Interim Design Standard Version 1.0*. London: AIG.
Arquati, D. 2008. *Pedestrians in central London lost and found: the Legible London wayfinding system* London: Transport for London. Available at http://www.eltis.org/sites/eltis/files/case-studies/documents/pedestrians_in_central_london_5.pdf, accessed 06/04/16.

Bennett, L, 2011. Bunkerology – a case study in the theory and practice of urban exploration. *Environment and Planning D: Society and Space* 29: 421–434.

Bonnett, A. 1989. Situationism, geography, and poststructuralism. *Environment and Planning D: Society and Space* 7: 131–146.

Bozatli, M., Lim, E.R. and Lim, R. 2004. *Downtown wayfinding: a study.* New York: Columbia University.

Burns, P.C. 1998. Wayfinding errors while driving. *Journal of Environmental Psychology* 18: 209–217.

Clark, S. 2008. *An analysis of pedestrian wayfinding systems and whether such systems can encourage greater levels of walking.* Dissertation submitted for a Masters in Built Environment Studies, London South Bank University.

Colin Buchanan 2008. *Legible London Bond Street prototype evaluation framework.* London: Colin Buchanan.

Cresswell, T. 2006. *On the move: Mobility in the modern western world.* London: Routledge.

de Certeau, M. 1984. *The practice of everyday life.* Berkeley: University of California Press.

Department for Transport, WalkEngland. 2008. *Walking maps.* London: Department for Transport.

Dodge, M., Kitchin, R. and Perkins, C. (Eds) 2011. *The map reader: theories of mapping practice and cartographic representation.* London: Wiley.

Fendley, T. 2015. What's next for legible cities? Available at https://segd.org/what-next-legible-cities, accessed 15/06/16.

Garrett, B. 2014. Undertaking recreational trespass: urban exploration and infiltration. *Transactions of the Institute of British Geographers*, 39: 1–13.

Golledge, R.G. 1992. Place recognition and wayfinding: making sense of space. *Geoforum*: 23: 199–214.

Greater London Authority 2013. The Mayor's vision for cycling in London. Greater London Authority. Available at: http://content.tfl.gov.uk/gla-mayors-cycle-vision-2013.pdf, accessed 29/05/16.

Greater London Authority 2016. Greater London Authority (GLA) Datastore. Available at: http://data.london.gov.uk/, accessed 20/06/16.

Harley, J.B 1989. Deconstructing the map. *Cartographica*, 26: 1–20.

Ichioka, S., Church, D. and Haider, S. 2005. *Legible London: Leicester Square navigation survey*, 11 March 2005. Prepared for Applied Information Group by Enterprise LSE Cities Ltd, London: London School of Economics.

Lynch, K. 1960. *The image of the city.* Cambridge MA, MIT Press.

Passini, R. 1981. Wayfinding: a conceptual framework. *Urban Ecology*: 5: 17–31.

Passini, R. 1996. Wayfinding design: logic, application and some thoughts on universality. *Design Studies* 17: 319–331.

Pinder, D. 2004. *Visions of the city: utopianism, power and politics in twentieth-century urbanism.* Edinburgh: Edinburgh University Press.

Pinder, D. 2013. Dis-locative arts: mobile media and the politics of global positioning. *Continuum*, 27: 523–541.

Schmitz, S. 1999. Gender differences in acquisition of environmental knowledge related to wayfinding behaviour, spatial anxiety and self-estimated environmental competencies. *Sex Roles* 41: 71–93.

Spinney, J. 2016. Fixing mobility in the neoliberal city: cycling policy and practice in London as a mode of political–economic and biopolitical governance. *Annals of the American Association of Geographers* 106: 450–458.

Transport for London 2004. *Making London a walkable city: the walking plan for London.* London: Transport for London.

Transport for London 2014. *Legible London evaluation 2013/14 report.* Steer Davis Gleave for TfL. London: TfL. Available at: http://content.tfl.gov.uk/legible-london-evaluation-summary.pdf, accessed 16/06/16.

Transport Research Laboratory (TRL) 2008. *PERS Legibility – Bond Street after study final report.* Guildford: TRL.

Urry, J. 2007. *Mobilities.* Cambridge: Polity.

Xia, J.C., Arrowsmith, C., Jackson, M. and Cartwright, W. 2008. The wayfinding process relationships between decision making and landmark utility. *Tourism Management* 29: 445–457.

4 Bicycle design history and systems of mobility

Peter Cox

Introduction

The bicycle is a strange object. It recurs with increasing frequency not only in today's considerations of mobility and sustainability, but as an illustrative example in social theory of the relations between objects, environments and subject identities. In the abstract consideration of the theoretical bicycle, however, the specific object, especially its multiplicity and diversity, can become obscured. Hadland and Lessing's (2014) authoritative work on bicycle design history succeeds by explicitly concentrating on the technological, not social, dimensions of the object, yet even they are forced to occasionally stray into wider consideration of the environmental, social, economic and political forces that have interacted to shape this history. This chapter focuses on the ways in which bicycle design connects with a range of factors; how external forces may shape reinterpretations of bicycle design, and how bicycle design, in turn, may be used to try to shape the external world.

Two historical cases are explored to show how bicycles, as design objects, are entangled with practices and identities: Germany in the 1920s and 1930s and England in the 1960s and 1970s. In the first case, design is used to reproduce and reinforce a dominant political ideology through reinterpretation rather than innovation. Here the bicycle allows new connections to be made between state and citizen. In the second case, design innovation is employed to challenge dominant ideologies of mobility: bicycles are used to connect citizens to new mobility practices. Both cases illustrate the relations between design and politics and both have implications for inclusion and access aspects of social justice. Both studies make use of close reading of manufacturers' literature but place it more strongly in a political/cultural context to understand the relationship between the design objects and wider society. The following section considers broader theoretical perspectives within which these case studies are situated.

Theorising bicycles as design objects

Thinking about the bicycle as a design object immediately confronts us with contradictions. The first is that most bicycle production, historically speaking, is not actually the result of deliberate design processes. A bicycle is an assemblage of a series of interrelated modular subsystems, many of which can be further broken down. For example, wheels are assembled from hubs, spokes and rims, then shod with tyres. A hub containing a brake and gear system removes the need for these functions as independent systems but necessitates changes in the interaction of the user and the machine, as well as making specific demands on the assembly and engineering of the wheel. To focus on bicycles as complete objects can be misleading. The bicycle, as a largely modular technology, is assembled as the interplay of separately conceived and designed subsystems and therefore almost always a compromise brought about by the availability, cost and accessibility of existing products, components and manufacturing processes, constrained by tradition and pre-existing social and embodied practices.

A central concern of this study, echoing that of Tonkiss is to consider the "social practices and processes that shape spatial forms, relationships and outcomes in intentional as well as in less intended ways" (2013, 5). These practices are themselves shaped by the affordances of the technologies through which they are pursued, and in the broader socio-technical contexts in which practices are performed (Kimbell 2012). In this context, we can frame these discussions within Yaneva's (2009) arguments for "design as a type of connector" refusing to separate the aesthetic and the technical, and to insist that artefacts are not simply 'things' but nodes of linkages in networks, of action at a range of levels. So, while the discussion may appear to focus on discrete objects, such as bicycles, and on the sales materials produced to accompany them, these are but elements in a broader sociography (Akrich 1992, 205).

Against this background, both Fallan (2008) and Yaneva (2009) demonstrate the utility of Actor Network Theory (ANT) and script analysis (Akrich 1992; Akrich and Latour 1992) in the study of design history. It is important to note that although scripts embrace "the vision of the world incorporated in the object and the program of action it is intended to accomplish" (Yaneva 2009, 284), users can decide not to follow the roles prescribed, or even to (wilfully) "misunderstand, ignore, discard or reject" scripted roles, opening up a division between spheres of production and those of consumption and use (Fallan 2008, 63). To read these scripts and to expose the manner in which they are constructed, this chapter undertakes a close reading of bicycles and the advertising material through which they are presented to the world. The examples are taken from two very different contexts precisely in order to reveal how bicycles interact with the broader worlds of which they are parts.

Existing literatures have examined bicycle design in social context relatively extensively. Rosen's pioneering (2002a) essay on the appropriation of the mountain bicycle highlighted the vital role of users in re-creating technologies. Other authors have followed this work by examining the relationships between bicycle technologies and identity formation and the frequently recursive nature of these in advanced industrial contexts (Huybers-Withers and Livingston 2010; Edwards and Corte 2010). While this approach can be helpful, more pertinent to the approach taken here is Marie Kåstrup's examination of how cycling practices have been utilised in the presentation of Danish national identity, where cycling by all groups in society is perceived to "correspond to an intrinsic Danish democratic spirit" (2009, 8). Neither the understanding of cycling nor Demark as democratic is fixed or inevitable, she argues, but both are constructed, and have been mutually constituted in support of changing political agendas for more than a century. In the examples here, discourses of the bicycle as equalising are also mobilised but in support of the efficient state rather than democracy.

Returning to the link between bicycle design and the self, Fallan (2013) explored the role of the Norwegian-made Kombi-bike as a source of memory and identity. In turn, Fallan (2013) provides a neat illustration of a point made by Illich in his analysis of the broader function of bicycles in a world of tools and technologies: "Tools are intrinsic to social relationships. An individual relates himself in action to his society through the use of tools that he actively masters, or by which he is passively acted upon" (Illich 1973, 24). For Illich, a transport system based on human-powered mobility (bicycles, handcarts, rickshaws and its corresponding road system together with repair stations) epitomised the capacity of hand tools to move from simple objects to constituting a complex system. What was important about the mobility provided by human-powered vehicles of all types was that he understood it to be self-limiting, that is, it

> creates only those demands which it can also satisfy. Every increase in motorized speed creates new demands on space and time. [The bicycle] . . . allows people to create a new relationship between their life-space and their lifetime, between their territory and the pulse of their being, without destroying their inherited balance.
>
> (Illich 1974, 75)

Through Illich's (1974) analysis of the bicycle as a tool, we observe that not only is it a repository of values, memories and social relations, but as a mobile object it has a different relationship to its demands on space than those reliant on external power sources. It is a tool for traversing and encompassing space, but the rate at which it can do so is limited by its status as a hand tool, functioning as an extension of the person. This emphasis on the particular spatiality of the bicycle in general may need to be modified when we come to consider the widely varied design history of bicycles, but it does provide a useful basis for considering the linkage between the bicycle as a design object and its spatiality.

To take the discussion further, we need to consider the relationship between tool and user. Colebrook (2002) uses the example of the bicycle to illustrate the difference between machine and mechanism. The latter is "a closed machine with a specific function" whereas a bicycle is an example of a machine that has no specific meaning: it only has meaning as it is connected with (an)other machine(s). "The human body becomes a cyclist in connecting with the machine, the bicycle becomes a vehicle" (2002, 56). If it is connected with another machine, such as a gallery wall, it becomes something else: an artwork.

In their connected form, machines become machinic assemblages: both cyclist and vehicle are assemblages. Malins' (2004) exploration of the potential of

Deleuze and Guattari as a means to understand these assemblages, suggests the cyclist, vehicle, bicycle system as one example in the following adapted excerpt (where [-] is employed to signify the assemblage):

> We will never ask what a [-] means, as signified or signifier; we will not look for anything to understand in it. We will ask what it functions with, in connection with what other things it does or does not transmit intensities, in which other multiplicities its own are inserted and metamorphosed, and with what bodies without organs it makes its own converge.
>
> (Deleuze and Guattari 1988, 4)

For the purposes of discussion here, the significance of a Deleuzian perspective on the machinic assemblage is that it problematises Illich's (1974) object orientation (which threatens to dissolve into attributing inherent qualities to specific tools) whilst allowing us to retain the value of his observation on the role of tools in social relations.

In sum, bicycles are tools that gain meaning as components of broader assemblages. In these assemblages they act as nodes in a wider series of connections. Bicycles as tools and design objects also convey 'idealised' scripts, connecting them to the political and ideological. The mobility of the cyclist consequently entangles physical objects, mobilities and semiotics in complex layers. Focusing on specific examples allows us to elucidate how design for mobility can play out in different ways dependent on the strengths and weaknesses of its connections and of the forces to which it connects. The case studies do not present exhaustive histories but focus on specific elements. It should also be noted that the conventional bicycle provides a very limited canvas for the designer, and for the communication of ideas and the construction of images. Cues are generally subtle. Changes in detail, perhaps unobservable from a distance, or by those unfamiliar with bicycle technologies, can communicate significant differences in status. Studying the broader canvas of the interplay of cycling technologies, advertising and contexts allows us to understand the ways that meanings of mobility objects and the practices associated with them are created.

From Bauhaus modernity to the national socialist bicycle

The utility bicycle in 1920s Germany was an established mass-manufactured and mass-market machine. Its success relied not on novelty but on images of reliability and familiarity. Individual firms maintained a wide diversity of engineering standards for more invisible elements of bicycles (such as the bottom brackets and cranks), but for those visible parts (which would be identifiable to the average purchaser) conformity to a recognisable commonality was an important factor. Utility cycles in Germany through the majority of the twentieth century were dominated by coaster brake hubs built into strong rear wheels. From before the First World War, these hubs obviated the need for better front brakes and so a simple plunger was retained as a standard fitting. Consequently, a basic frame, diamond for men, twin curved tubes (pre-shaped by the tube manufacturers) for women, with coaster hub and plunger-type front brake, remained a standard and rarely varied formula for over half a century, long after rim brakes had become the norm elsewhere in Europe.

This stabilisation and standardisation of design framed the bicycle as a product of industrial (not craft) production. Coaster brakes require relatively heavily built wheels, and thus a bicycle built to these standards provides a different script to (for example) a British (hand-crafted) lightweight bicycle of the same period. To use Akrich and Latour's (1992) definition, the specificity of production standards in specific territories prescribes and proscribes significantly differing functions. The point to emphasise from this discussion is that as a design object, the German bicycle of the 1920s and early 1930s was stable to the point of stasis: minimal changes distinguished utility bicycles from year to year. How the bike, its rider and riding as an activity were presented to shape the meaning of design thus becomes of greater significance than any design changes. In order to understand this we need to look at the socio-technical scripting of the bicycle rather than its practical affordances.

The 1925 catalogue of bicycle manufacturers Burgsmüller & Sohne of Kreiensen am Harz (FS5502199) provides an example of the way in which makers tried to reposition the bicycle in the world of utility consumer goods without redesigning it. In other words the qualities of the bicycle as a design product are moved or changed not through significant alteration to the object itself, but through making novel connections with other, external factors. These connections reframe the meanings of the object as much as any intrinsic qualities of the objects themselves. The catalogue strikingly combines both tradition and modernity in its presentation of the bicycle. For its front cover, it illustrates the development of the bicycle "from the Draisine, through the Highwheeler, to the modern bicycle", in a visual style that is distinctly modern in comparison with other catalogues of the period. Colour and general layout pay homage to stylistic cues emerging from the Bauhaus. Typography, whilst not as extreme as that found in expressionist cinema, is strikingly modern and at odds with the dominant use of Gothic in other contemporary sales catalogues (and to which it returns in interior pages). These connections position the bicycle, by implication, as a truly modern product, without significant change to the object itself.

The interior layout of the journal, by contrast, is mostly conservative, in keeping with the majority of the range of cycles produced. Where the modern style is free to emerge in relation to the bicycle itself is in their single, colour-illustrated, racing model. Here bold colour is used on the frame and even red tyres are fitted to mould a strikingly different image of the bicycle to the monochrome (usually black) machines in the rest of the catalogue. For all the rest of their output, however, existing conventions are unbroken, including the use of stylistic conventions common in bicycle design from the turn of the century onwards. Luxury, for example, is often indicated by the use of box lining (as might be found in the detailing of furniture). Double box lining characterises the most expensive variants in the Burgsmüller catalogue. Indeed, the visual layout and presentation are not dissimilar to the style details in Biedermeier furniture. What we see is therefore a bifurcation between the scripts of the sports bicycle and the bicycle as everyday consumer item. However, the two are held together in a creative tension, combined within the same catalogue pages. The unchanged models are reframed as modern by association through chains of connection: not just to the racing model, but also in relation to Bauhaus imagery.

By framing the whole in a narrative of historical progression yet drawing visual cues from a more conservative direction, the Burgsmüller catalogue positions the bicycle as a strikingly modern product, but simultaneously one with a trustworthy heritage. If the machine is not itself symbolic of modernity, as it might have been 30 years earlier, the inclusion of innovatively styled models that connect with other tropes of modernism shows the reader and prospective purchaser that it can be so. Excepting the cover, riders are rendered invisible; emphasis is entirely on the machine itself.

Moving forward to the mid-1930s, the socio-political context of the bicycle in Germany was transformed. Emphasis on the high cultural value of the motor car had begun to grow in Weimar Germany, but under National Socialism it was formulated as state policy (despite Germany having one of the lowest rates of car ownership in northern Europe). Ebert sums the situation up:

> In 1935 an estimated 16 million cyclists faced 2.2 million cars in Germany. But in terms of traffic laws, the priorities lay with the minority. The new traffic law in 1934 stated that the promotion of the car was the highest goal ... the organisation of traffic was to serve this goal.
>
> (2004, 361)

The motor car promised both mass mobilisation and mass motorisation – fundamental to rearmament and economic reorganisation (Overy 1995). It was scripted as the quintessential vehicle (in both senses) of modernity. Cars however, were simply not to be realistically available to the majority population for the immediate future. One way of achieving state aims was to blur (and even remove) boundaries between bicycles, mopeds and motorcycles. Taxation and regulatory barriers were removed from low-power motorcycles and most bicycle manufacturers also provided motorised models, using widely available standard add-on engines.

The marginalisation of the bicycle in the ideological discourse of modernity and roads policy had to be tempered with the practical reality of the bicycle as the dominant mode of personal transport. For manufacturers and retailers, one way of doing this was to use the bicycle to display one's conformity to the regime. The formation of the state labour service organisation (*Reichsarbeitsdienst, RAD*) in 1935 and the formalisation of military sports (*Wehrsport*) as an integral part of training in the *hitlerjugend* provided manufacturers with an opportunity to market models specifically equipped for these paramilitary activities. Advertising sheets for Phänomen bicycles (Phänomen-Werke Gustav Hiller A.G., Zittau FS504602-504610) show a new model, the Wehrsportrad (Nr.20). Headlined "Der Güte Kamerad, für Wehrsport und arbeitsdienst", it is billed as the "abuse-resistant, robust, high-performance machine for on and off-road use". Advertising copy for 1934 proudly announces the Phänomen as "The bicycle of the German Championship" held by a model Aryan raising a fascist salute. Bicycles thus are configured as a means through which an idealised citizenship can be performed, maintaining relevance through recourse to national values if not modernity.

One should not, however, think that the bicycle trade was simply adapting to external political circumstances. Bicycles and cycling were actively used to assist the ideological construction of the national socialist regime. This is most clearly seen in the advertising material of the Torpedo-werke A.G. (FS504738-504744). In itself, the advertising copy is fairly unremarkable:

> The daily journey to the workplace is only a stone's throw away on a "Torpedo". And a bicycle tour on the weekend is a real pleasure, because, without exertion, tour locations, picturesque forests, quiet lakes are also within easy reach.

Here, reinscription of the bicycle comes about through the operations of broader social and political movements: the same copy—the literal script cited above— may take on very different aspects depending on context. In Britain, similar phrasing was used to promote the sense of social inclusion; freedoms are available to all, regardless of social class.

However, read against a backdrop of discourses of blood and soil and homeland in the Third Reich, the text of the Torpedo advertising copy takes on a different light (Stargardt 2015). In combination with the illustrations in the same brochure, these impressions are reinforced. One image (see Figure 4.1) presents two men riding towards a modern factory building. Contrasts in hat and jacket are sufficient to imply that this is both worker and manager side by side—an illustration of the elimination of class conflict, though again this could be read as an aspiration of egalitarian democracy. In context, however, this single illustration provides a striking example of what Tymkiw describes as

> contemporary rhetoric … framing the factory as both the former place of workers' alienation and a site, under national socialism, of workers rebirth and renewal … a seedbed from which a new 'people's community' would emerge.
>
> (2013, 368)

In case the context is unclear, immediately below is an image of a uniformed paramilitary.

Throughout the Torpedo advertising leaflet, the mobility of the bicycling citizen is connected to the ideological construction of citizenship, not only in respect of class identities, as the above example demonstrates, but also of gender roles. In another panel the text reads: "He and she are enthusiastic about *Torpedo*, because *Torpedo* manufacture not only a strong yet feather-light up-to-date man's bicycle, but also a functional ladies' bicycle, guaranteeing the highest ride-comfort". Masculinity is constructed as technologically modern, femininity as obedient and subservient, as the male rider points out the landscape features to his partner.

Figure 4.1 Torpedo Werke advertising leaflet, Germany, 1938.

Gender difference is emphasised. Additionally, the description of the delights of local touring echoes formulations that construct the landscape as an ideal part of national identity, experienced through mobility: the central ideal of early auto-bahn construction (Zeller 2007).

In these examples, there is little significant material innovation in the bicycle as object, yet elements of design are mobilised to create new meanings and a new assemblage through the connections made with ideological constructions of the citizen. The bicycle becomes a means through which citizenship can be con-formed to ideological norms. Design becomes politicised even without obvious visual clues (although these were also present through optional extras such as Hakenkreuz mudguard ornaments, FS592340).

The inseparability of bicycles and riders provided advertisers easy means through which to align aesthetic scripts in advertising materials with political ideology (Heskett 1978). More than that, these materials provide a scripted nar-rative of the ideals of citizenship roles and practices, suggesting that promoting bicycles and cycling as a consumer product can be a means of "management of consumption ... to temper mass desires without class conflict ... or capitulat-ing to 'materialism'" (Baranowski 2004, 39). While showing limited denotative innovation, both cycles and their advertising brochures recreate the connotative meanings of cycling, aligning them to an ideological construct of citizenship (Folkmann 2011). By connecting the bicycle into a different set of discourses—in this case of the obedient citizen—the meaning of the bicycle is shifted whilst the design object remains static. The mobility of the cyclist is no longer simply movement, but (re)constructed as ideological performance, embracing narratives of class and gender roles, the elimination of social conflict and identification with Homeland.

The Moulton bicycle in 1960s Britain

The second case study examines a point in time where radical changes to the design of the object were made to try and deliberately disconnect the bicycle and cycling with existing sets of connections and ideological assumptions, and to consciously seek new assemblages of practice.

By the mid-1950s, a revolution in the image of individual mobility was clearly under way. Italian manufacturers of motor scooters for example had set out "not just to make a new category of machines but a new category of consumer . . . and the conversion of consumption into lifestyle" (Hebdige 2004, 139). Designs were rapidly copied internationally and outstripped, even eclipsed, conventional light motorcycle sales. Not only were scooters clearly different in form because of their small wheels, but they were presented as a complete package, without visible engines or external mechanisms, unlike the traditional motorcycle with its frame and engine. Scooters created a new mobility aesthetic with mass appeal, especially in the aftermath of utilitarian austerity across all the European nations recovering from the war. Here changes in design were used to change the meaning of the object.

Clearly inspired by scooter design and the transformations with which they were associated, engineer Alex Moulton undertook a radical reconsideration of bicycle design (Hadland 2000, Moulton 2009). Small wheels (16 inches as opposed to the 27-inch diameter of a standard bicycle wheel) and integral suspension marked this new design as fundamentally different from any previous mass-market bicycle. Coming on to the market at the end of 1962, the simple lines of its "step through"' frame and integral luggage racks made a strong design statement. The Moulton range of bicycles was self-consciously unlike the conventional bicycle: it provided the opportunity for a new range of scripts to be written, by producers and by users.

Not only was the look of the machine strikingly novel, so too was the advertising copy used to market it (see Figure 4.2). Drawings of riders and environments disappeared; instead a bold silhouette was used to highlight the radically different profile of the machine, sometimes overlaid on an image of a conventional large-wheeled bicycle to emphasise its difference. A keyword used was "Universal"; emphasis laid on the fact that the same frame could be used for a range of purposes, with slightly different equipment options for touring, everyday use and even for racing. Likewise, there was only one frame size, intended to fit a majority of the population with instantaneous adjustment through quick-release fittings. Neither was the bicycle gendered in conventional ways: this was a "unisex" design to employ the word coined in the mid-1960s to epitomise contemporary questions about the appropriateness of existing roles and gender identities.

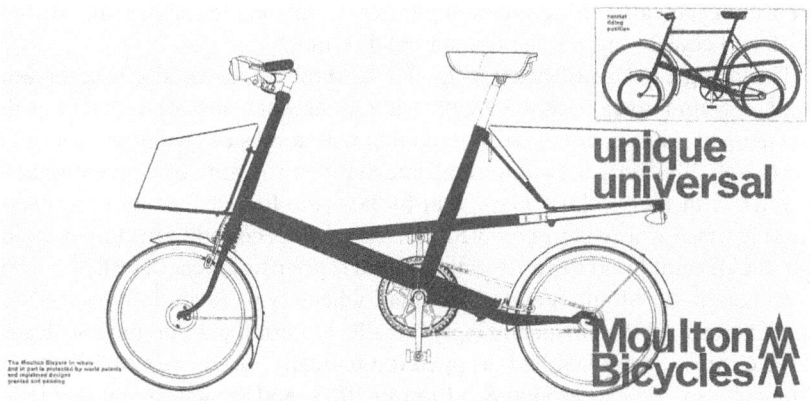

Figure 4.2 Moulton bicycle advertising leaflet, UK, 1963.

This universality can also be read as a means to break away from issues of social class dominating public perceptions of post-war utility cycling in the UK. As Epperson (2013) argues, the success of Moulton was in engaging customers who had no prior intention of purchasing a bicycle before they encountered the Moulton. It appealed to new users precisely because it was not a 'proper' bicycle with all the cultural baggage that had been collected in the past half century. Here again we see a classic case of re-inscription but one that attempts to sever connections to the existing socio-technical script through re-imagining and re-working aesthetics through design. These same factors that made it acceptable to new audiences also alienated many existing bicycle users who rejected it precisely because it was not seen as a 'proper' bicycle.

The Moulton sold well initially, and sales success prompted other manufacturers to bring out their own small-wheeled machines. Raleigh, with 75 per cent of the existing bicycle market, spent unprecedented sums on its own advertising campaigns and, importantly, focused on new target audiences. Popular general circulation magazines and 'women's interest' titles provided previously unconsidered opportunities (Hadland 2011). Indeed, Raleigh's market research pointed to an "unusually high percentage" of women purchasing or riding Moulton and other small-wheeled bicycles (Epperson 2013, 243). The core of success of the Moulton and its successors in the UK reflected a re-imagining of the bicycle as "consumer goods, not bits of light engineering" (Peter Seales, head of Raleigh Marketing in 1973, cited in Rosen 2002b, 102).

The universality scripted into the design of the Moulton in the early 1960s can be seen to engage not only with rethinking of physical mobility, but also of social mobility. The design itself was not prescriptive of specific uses or users, but offered a flexible platform that could be interpreted in many different ways. The same frame was common across the entire range of production, from basic single-speed models, through touring variants to racing models. This also challenged the tradition that the 'best quality' bicycle should necessarily be bespoke. Moulton did offer a specialist 'S' range of super-deluxe models, but these only differed in finish and equipment, not the fundamental frame.

The design not only fitted with the cultural tropes of the scooter and the 'mini'—bound up as they were with other emergent issues of age, and of design as statement—but in doing so, layered this with issues of changing class identity. Conversely, this only served to reinforce the identification of conventional urban bicycles with the working class, and by now the lowest income sections of the working class who were perceived as lacking the economic resources to join in with the general trend towards motorisation (Oldenziel *et al*. 2016).

Although the Moulton and other small wheelers were a relative success, they were launched into an increasingly hostile environment for bicycle transport. Bicycle traffic in the UK had dropped dramatically, from over 20 billion vehicle kilometres in 1954 to around 8 billion in 1963, and growth in car use was promoted as the primary aim of transport policy and urban planning. Conservative transport minister Ernest Marples summed up his agenda in 1959, stating that "we have to rebuild our cities. We have to come to terms with the car" (Hamer 1987, 54). The bicycle afforded different uses and users, but the context in which it was ridden afforded fewer and fewer possibilities for use.

While this new constellation of designs provided new scripts for the bicycle and enabled new connections of riders, technologies and meanings, the technologies themselves were not capable of more than temporarily halting a general decline in bicycle use. While advertising imagery suggested these small-wheeled bicycles as iconic of a new urban chic, users were not easily able to enact this script, however much they may have desired it, because of external factors that increasingly facilitated and prioritised provision for mass motorisation, competing unequally for urban space. Moreover, large numbers of existing cyclists rejected this new potential set of meanings as irrelevant to, or even threatening of, their existing practices (according to extensive correspondence in the CTC Gazette 1963–1968). To turn once more to the vocabulary of Akrich and Latour (1992), we see responses to the novel script provided by the Moulton as simultaneously both subscription and de-inscription. The new affordances are underwritten by some, while others deny or refute them, even seeing them as an additional threat to their already threatened identities and practices. Producers' intentions may well have been to open up the potentials of urban cycling to new groups of consumers and in many ways they did just that. Hadland (2012) argues convincingly that subsequent small-wheeled bikes reshaped the concept of the bicycle for much of the British public. However, the mobility environment that they then entered was structurally incapable of supporting their mobility. In Britain, government policy through the 1970s ensured the continued near invisibility of the bicycle as a means of transport (Cox 2015).

Conclusions

As mobility objects, bicycles script not only mobility practices but the mobility and mutability of subjectivities. To understand the processes of scripting, we need to look not only at the technologies themselves, but the much broader sets of connection in which they are entangled. In the 1920s and 1930s examples discussed in this chapter, design remained relatively static. Meanings, however, moved through the connections made with external factors, whether to design as a cultural landmark (Bauhaus) or to the ideology of fascism in the 1930s. The mobile subject in each case is constructed through these webs of connection, not only between designer/producers, technologies and users, but also with the environments in which they are used and the cultural discourses with which they are engaged. In fascist Germany, the bicycle—as a design object—became a means through which the mobile subject could perform a political identity, ensnared with particular constructions of gender and class roles. The mobile subject is itself 'designed' through the scripts provided by the bicycles as objects and their sales materials. Although the physical design of the bicycle is relatively static, the manner in which it is mobilised through connection to other tropes creates the user as a political subject and legitimates specific mobility practices as part of that political subjectivity.

In the 1960s, Alex Moulton consciously set out to confront the stasis of bicycle design and the determinate scripts accrued over time and associated with the bicycle as people recognised it. Breaking with iterative design changes, the resultant machine succeeded in gaining freedom from prior technical and socio-technical scripts. In doing so it made significant market in-roads and opened the door for numerous subsequent changes and rediscoveries of the bicycle as fashionable and acceptable later in the 1970s (Watson and Grey 1978). Design innovation did create (temporarily at least) new mobilities thinking and arguably paved the way for many later re-imaginings of the place of the bicycle in the city. Ultimately, however, its ability to reinvent social relations was circumscribed by externalities that were more powerful. The recovery of cycling as an everyday practice in a number of cities around Europe in the late 1970s was not the direct product of innovative design thinking but of rearrangements in urban politics.

The cases presented in the chapter highlight the importance of connecting concerns with design objects with the complex connections in which they are entangled. Both studies could easily be extended into much broader webs of interconnectivity, revealing other dimensions of interplay, in which the scripting of objects shapes user performances and these in turn de-script and re-inscript objects with new meanings. The bicycle is one node in these connections through which mobilities are designed and scripted. To achieve greater understanding we need to expose the extent of configurations within greater assemblages.

Note

Acknowledgement (and many thanks) is given to the Deutsches Museum, Munich, whose archivists were of immense help in tracking down the material, and to the Leverhulme Trust, for funding the International Academic Fellowship (IAF-2014-016) which provided the time needed to engage in this research.

References

Akrich, M. 1992. The de-scription of technical objects. In *Shaping technology/building society: studies in sociotechnical change*. Ed. W.E. Bijker and J. Law. Cambridge, MA: MIT Press, pp. 205–224.

Akrich, M. and Latour, B. 1992. A summary of a convenient vocabulary for the semiotics of human and nonhuman assemblies. In *Shaping technology/building society: studies in sociotechnical change*. Ed. W. E. Bijker and J. Law. Cambridge, MA: MIT Press, pp. 259–264.

Baranowski, S. 2004. *Strength through joy: consumerism and mass tourism in the Third Reich*. Cambridge: Cambridge University Press.

Colebrook, C. 2002. *Understanding Deleuze*. Crows Nest, NSW: Allen and Unwin.

Cox, P. 2015. Rethinking bicycle histories. In *The invisible bicycle: new insights into bicycle history*. Ed. T. Männistö-Funk and T. Myllyntaus. Leiden, Netherlands: Brill.

Deleuze, G. and Guattari, F. 1988. *A thousand plateaus: capitalism and schizophrenia*. London: The Athlone Press.

Ebert, A-K. 2004. Cycling towards the nation: The use of the bicycle in Germany and the Netherlands, 1880–1940. *European Review of History* 11: 347–364.

Edwards, B. and Corte, U. 2010. Commercialization and lifestyle sport: lessons from 20 years of freestyle BMX in 'Pro-Town, USA. *Sport in Society* 13: 1135–1151.

Epperson, B. 2013. A new class of cyclists: Banham's bicycle and the two-wheeled world it didn't create. *Mobilities* 8: 238–251.

Fallan, K. 2008. De-scribing design: appropriating script analysis to design history. *Design Issues* 24: 61–75.

Fallan, K. 2013. Kombi-nation: mini bicycles as moving memories. *Journal of Design History* 26: 65–85.

Folkmann, M.N. 2011. Encoding symbolism: immateriality and possibility in design. *Design and Culture* 3: 51–74.

Hadland, T. 2000. *The Moulton bicycle* (third edition). Coventry: T. Hadland.

Hadland, T. 2011. *Raleigh: past and presence of an iconic bicycle brand*. San Francisco: Cycle Publishing / Van der Plas Publications.

Hadland, T. 2012. *Raleigh Twenty (R20)*. Available at https://hadland.wordpress.com/2012/06/24/raleigh-twenty-r20/, accessed 4/3/2015.

Hadland, T. and Lessing, H.E. 2014. *Bicycle design: an illustrated history*. Cambridge, MA: MIT Press.

Hamer, M. 1987. *Wheels within wheels: a study of the road lobby*. London: John Murray.

Hebdige, D. 2004. Object as image: the Italian scooter cycle. In *Material Culture*. Ed. V. Buchli. London: Routledge, pp. 121–160.

Heskett, R. 1978. Art and design in Nazi Germany. *History Workshop* 6: 139–153.

Huybers-Withers, S.M. and Livingston, L.A. 2010. Mountain biking is for men: consumption practices and identity portrayed by a niche magazine. *Sport in Society* 13: 1204–1222.

Illich, I. 1973. *Tools for conviviality*. London: Calder and Boyars.

Illich, I. 1974. *Energy and equity*. London: Calder and Boyars.

Kåstrup, M. 2009. Indentity and bicycle culture – a Danish perspective. Presented to Cycling and Society Research Group Symposium. Bolton, September 2009.

Kimbell, L. 2012. Rethinking design thinking: Part II. *Design and Culture* 4: 129–148.

Malins, P. 2004. Machinic assemblages: Deleuze, Guattari and an ethico-aesthetics of drug use. *Janus Head* 7: 84–104.

Moulton, A. 2009. *Alex Moulton: from Bristol to Bradford-on-Avon – a lifetime in engineering*. Derby: The Rolls Royce Heritage Trust.

Oldenziel, R., Emanuel, M., de la Bruhèze, A.A. and Veraart, F. 2016. *Cycling cities: the European experience*. Eindhoven: Foundation for the History of Technology.

Overy, R.J. 1995. *War and economy in the Third Reich*. Oxford: Clarendon Press.

Rosen, P. 2002a. Up the vélorution: Appropriating the bicycle and the politics of technology. In *Appropriating technology*. Ed. R. Eglash, J. Bleecker, J. Croissant, R. Fouché and G. Di Chiro. Minneapolis: University of Minnesota Press.

Rosen, P. 2002b. *Framing production: technology, culture and change in the British bicycle industry*. Cambridge MA: MIT Press.

Stargardt, N. 2012. *The German war: a nation under arms, 1939–45*. London: Bodley Head.

Tonkiss, F. 2013. *Cities by design: the social life of urban form*. Cambridge: Polity Press.

Tymkiw, M. 2013. Art to the worker! National Socialist Fabrikaustellung, slippery household goods and Volksgemeinschaft. *Journal of Design History* 26: 362–380.

Watson, R. and Grey, M. 1978. *The Penguin book of the bicycle*. Harmondsworth: Penguin.

Yaneva, A. 2009. Making the social hold: towards an actor-network theory of design. *Design and Culture* 1: 273–288.

Zeller, T. 2007. *Driving Germany: the landscape of the German autobahn, 1930–1970*. Oxford: Berghahn.

5   Rushing, dashing, scrambling

The role of the train station in producing the reluctant runner

Simon Cook

Introduction

This chapter forms part of postdoctoral research attempting to rethink contemporary running mobilities in the UK. In this research project I have focused on two kinds of running: first is the growing number of people who choose running as their mode of commuting. This represents a purposeful, intentional and planned form of running as transport (Cook, 2016) and is commonly associated with wider goals of sport, fitness and health (Cook *et al.*, 2015). These associations are quite recent, the increasing ubiquity and normalisation of running not truly occurring until the running boom of the 1960s and 1970s (Latham, 2015).

My interest here, however, is to explore other more mundane and less structured instances of running in contemporary society. Hence the second kind of running as transport I explored were the unintended, improvised and often undesirable moments of rushing, dashing and scrambling that are common in everyday life. Pause long enough at any road crossing, bus stop or train station and the sight of people dashing, in what might be considered an everyday emergency mobility, is something you will witness soon enough. These are undoubtedly little emergencies, where something has gone wrong (usually concerning something taking longer than anticipated) and running is unexpectedly employed in order to avoid something undesirable (such as missing your train). Unlike more serious emergencies, dangers here are not life threatening (though the consequences of being late can be serious, like losing your job) and are generally resolved by moving towards rather than away from something. So whilst more individual and small-scale than forms of emergency mobility explored hitherto in the literature (Adey, 2016), analysis of these everyday emergency forms of mobility have much to tell us about mobility as an accomplishment.

It is the aim of this chapter to detail and analyse instances of 'emergency' running in the train station, paying particular attention to the ways in which mobilities are produced through the coming together of figurative and physical design features. The chapter begins with a general discussion of the design of train stations and running spaces before introducing the case study (Guildford train station in South East England). The main discussion focuses on three aspects of station design— temporal staging, semiotics and the material site—which affect the production of emergency-running through their presences and absences. These will be returned to in the conclusion, where their importance in producing certainty and uncertainty will be discussed.

Station design and running

Like other mobility hubs, a contemporary train station is designed with many activities and ideas in mind (Jensen, 2014). They function much beyond the simple embarking and alighting of rail passengers, though they are carefully designed to facilitate such flows. Train stations are also spaces of administration, of safety and security, of commerce and capital, of relaxation and sociality, as well as boredom and suspension (Bissell, 2007). They are spaces designed to be inhabited by different bodies and navigated by forms of mobility that harmonise this varied functionality and patronage. In particular, train stations specifically produce slowed mobilities of walking and waiting.

Running is not one of the range of movements or activities that station designers consider to be likely or appropriate. Mostly out of concerns for passenger safety, the design of train stations tends to demonstrate an attempt to discourage and inhibit running in these spaces. Most certainly, train stations offer none of the design specialisations that seek to make running easier or more desirable (see below). Yet running does occur abundantly in train stations as a form of everyday emergency mobility. This running is often unintended and improvised but occurs when people lack other means with which to cover a particular spatial distance within particular temporal restrictions, as in this case, needing to run to catch a departing train.

The improvised nature of this emergency-running demonstrates an interesting relationship between design, bodies and mobility. It not only reveals insights into encounters between people and design and the ways design affordances produce particular space and mobilities, but it also asks how appropriating spaces not specifically designed for running changes how we understand running as a mobile practice. Ultimately, this relates back to the differing role running takes on in these scenarios compared with the dominant associations of running with sport and fitness.

The design of running spaces

While not necessarily apparent in instances of dashing and rushing, running more generally is a highly designed practice. Hegemonic ideas about what running is and the role it should play in society feed into the practical design of running spaces, variously seeking to optimise either the sporting or fitness aspects of running by facilitating an improvement in speed or by encouraging participation and enhancing experiences.

Take for example the competitive running space of the athletics track, an unmistakably designed sportscape of rationality, artifice and standardisation. With its regulated spatial parameters and continually refined surface, the athletics track is a synthetic and technological monoculture designed to offer segregated, controlled, neutralised and predictable conditions for optimal sporting performance, replicability and comparability (Bale, 1994; 2003). Fitness running spaces on the other hand, have tended to be designed with a concern for enhancing running's health benefits. For example, Swedish jogging tracks are designed to improve aerobic capacities and take advantage of the restorative role of nature by being built in forests with undulating terrain (Qvistrŏm, 2013; 2016). Other designs seek to improve participation and motivation in mass fitness running by experimenting with surface materials, wayfinding and pacing technologies among many other things, such as the case of bark running tracks in Flanders, Belgium (Borgers *et al.*, 2016). These noted, the majority of running in England actually takes place in spaces not directly designed for it—the shared spaces of public streets and parks (Sport England, 2012). Runners' appropriations of these unspecialised spaces, whether individually (Cook *et al.*, 2016), as part of mass organised events (Cidell, 2014), or through digital activity tracking, mapping and sharing (Carléna and Maivorsdotter, 2016), is productive of unique social relations and micro-politics.

Running in a train station, however, lacks almost all of these specialisations and design ideals for encouraging participation and improving performances/ experiences. It is a very different type of running and a very different experience of running to those most commonly considered. This chapter seeks to account for these differences by exploring the role design has in producing the types of running seen in a train station.

The case study: Guildford station

The case study upon which this ethnographic research is based is Guildford station in Surrey, a highly affluent town of around 140,000 inhabitants in the south-east of England (Trott, 2015). Guildford's main railway station is managed by the company South West Trains, one of twenty rail franchises which operate on the UK national rail network. South West Trains has a Central London terminus at London Waterloo and provides the majority of commuter services to South West London, as well suburban and regional services to Hampshire, Berkshire and Surrey (as in this case). Guildford station acts as a major hub of mobility flows in the region, with its eight platforms drawing passengers in from and dispersing them to locations across England's south east, including London Gatwick Airport, Reading, Portsmouth and Central London. The station has around 8 million passenger entrances and exits per year (Office of Rail and Road, 2016). Guildford is the second most popular town in the South East for commuters into London. Although chiefly a space of rail mobilities, the station also serves as a transport interchange with bus stops, cycle parking and limited car parking surrounding the station. It is also a commercial space hosting shops at various locations throughout the station.

The ethnographic fieldwork took place over five consecutive weekdays in the summer of 2014. Permission to conduct the research was obtained from South

West Trains and a one-day pilot study was undertaken to identify critical sites and rhythms of emergency-running. Each day then consisted of spending eight hours observing passenger movements in the station. This was divided into three hours in the morning, from 06:30 to 09:30, two hours in the afternoon, from 12:00 to 14:00, and another three hours in the evening, 16:30 to 19:30, in order to observe the two peak commuting periods as well as the midday lull. I spent a full day in each of four chosen sites throughout the station, with a final day spent split between them (details below). This resulted in 1,286 counts of people running being made and 542 of those instances being noted down by hand. These were then typed up, coded for particular characteristics of emergency-running, and combined with sketches and site descriptions to ultimately steer the themes of analysis.

These in-field observations brought to light the immense importance of design to the production of emergency-running. It enabled the grasping of atmospheres, affects and the less-visible, multisensory elements of emergency-running, as well as the moments of transition, physical movements, body language and facial expressions. This led to a focus on three key aspects of the station design, which in their presences, absences and intersections played an important role structuring mobility: temporal staging, semiotics and the material site. These will be used to structure the main discussion.

The four different observation sites identified as particularly influential in the taking-place of emergency-running in the station were: (1) the Ticket Hall; (2) the Gateline; (3) the Underpass; and (4) Platform Five. Their locations in the station can be seen in Figure 5.1 and descriptions of each are given below.

The first site, the Ticket Hall is also the main entrance to the station (Figure 5.2). It is a complex site that must be negotiated in order to gain access to departing trains. Many of the functions of the train station intersect here, and as such, there are many obstacles, distractions and activities that require mobilities to be slowed and the flows of passengers to be sorted. The Ticket Hall is designed to be a space of inefficient dispersal, in order to facilitate the administrative, security and commercial functioning of the station. The rail-essentials of ticket-buying (from self-service machines or actual people), time-checking (on real-time digital displays or printed timetables) and ticket-validating (through the Gateline barriers or attendants) sit alongside distractions which encourage people to linger, wait, relax and spend, including coffee shops, food shops, newsagents and seats. It is thus a space of conflicting rhythms, where passengers for different trains, with different senses of haste, all intermingle. The platforms can just be seen from this location but no departure announcements can be heard in the Ticket Hall, so passengers tend to rely on the large display boards for train information.

The other three study sites are simpler in design and function. The Gateline acts as a bottleneck from the Ticket Hall. It is a linear space, guiding passengers through a security check onto Platform Two and providing access to the other seven platforms via an underpass or a bridge. It is designed with one solitary but significant obstruction to seamless passage—the ticket barrier. The six gates act to sort passengers with legitimate tickets from those without. In principle, their operation is simple—a passenger enters the gate and inserts their ticket, where it is

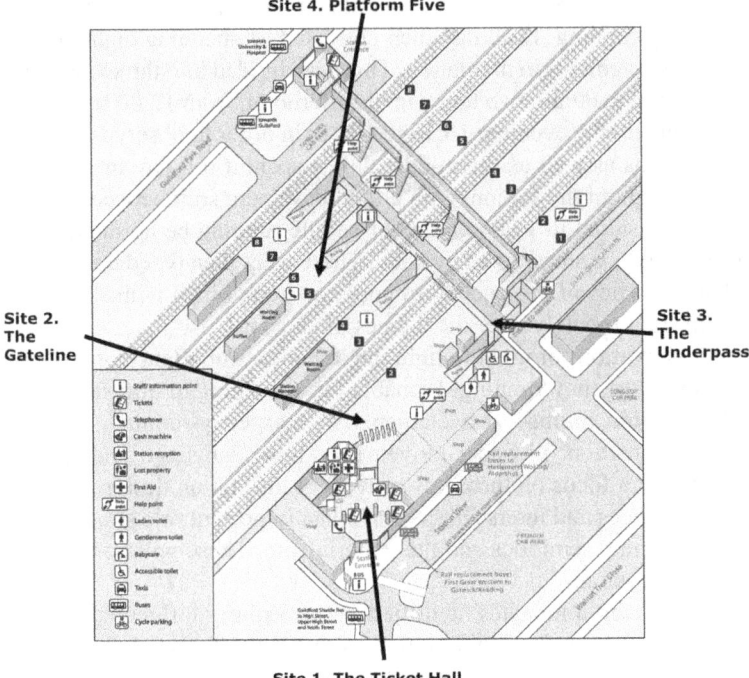

Site 4. Platform Five

Site 2.
The
Gateline

Site 3.
The
Underpass

Site 1. The Ticket Hall

Figure 5.1 Station plan showing observation sites.

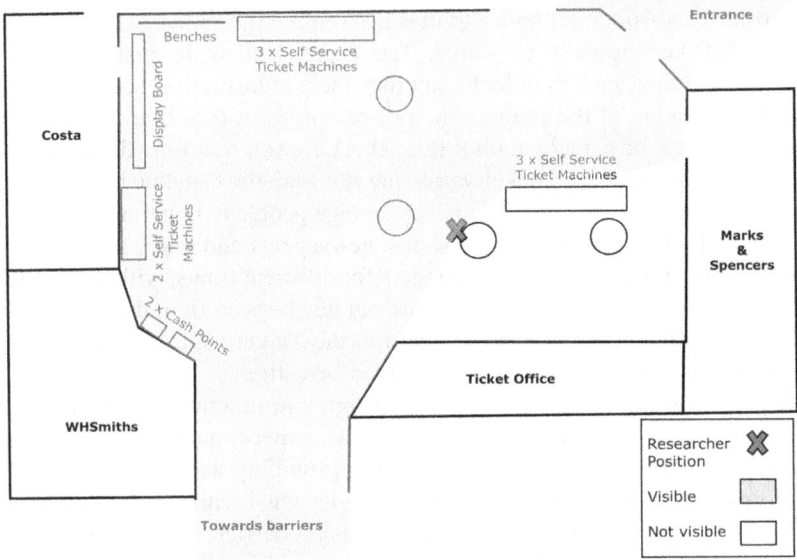

Figure 5.2 Plan of the ticket hall (not to scale).

electronically scanned and codes check its validity before permitting or restricting passage (they are only wide enough for one person, with the exception of the wide gate meant for wheelchair users and those with bikes, pushchairs and luggage). An attendant capable of manually opening any gate is also present in case issues arise. Once these are negotiated, passengers find themselves on Platform Two, again a linear space where, unless needing a train from this platform, passengers are channelled towards other platforms. Four electronic displays provide information for departures across the whole station, while another one displays detailed information about upcoming departures from Platform Two. These were mostly redundant, however, only being used in abnormal conditions such as delays or cancellations. While all platforms can be seen from here, only audio announcements for Platform Two can be heard.

The Underpass (Figure 5.3) is the more popular of the two means of accessing Platforms Three to Eight at Guildford station (the other being a raised walkway with steps down to each platform). It is a subterranean passage, entered by a downhill ramp whose entire function and design concern the efficient and speedy channelling of passenger flows to and from platforms. There are no places to wait or linger here. The artificially lit, cramped and dank main corridor only leads to uphill ramps to other platforms. Small display boards at either end provide the only indication of how 'on time' passengers are. All auditory cues of approaching trains are muffled and displaced, and platform-visibility completely removed.

Finally, Platform Five was chosen as it is the station's busiest platform. Only trains on the major trunk line between Portsmouth and London depart from here, offering the quickest transit to Central London. Arriving here on the uphill slope

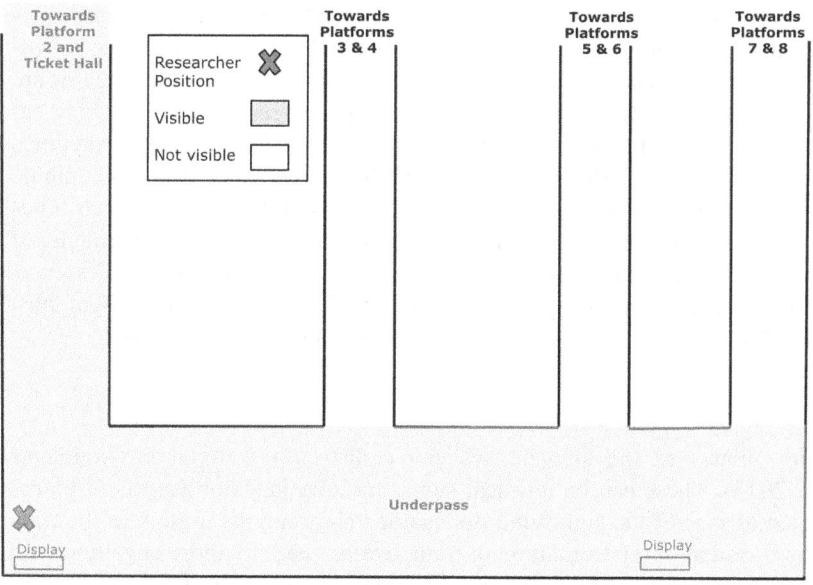

Figure 5.3 Plan of the underpass (not to scale).

from the seclusion of the Underpass, passengers are met with an open but covered platform with few permanent obstacles to an awaiting train. The platform disperses passengers linearly along its length as they sit, and wait for arriving trains. This can produce a space littered with undirected bodies and luggage. All semiotic cues return here: digital display boards provide information about the progression of forthcoming trains and auditory cues are fully perceptible.

Each site and its design affords something different in producing mobility in the station. Yet some intersecting aspects of this diverse design can be identified— temporal staging, semiotics, and material layout. The varied presence and absence of these different elements work to encourage and discourage emergency-running in different ways and helps to understand how mobilities are accomplished.

Affordances of design and the taking place of emergency-running

Temporal staging

The temporal staging of rail mobilities undoubtedly had the biggest effect upon emergency-running in this research. Simply, the design of train timetables creates the fixed common points in time by which passengers are judging their progression, and which dictates the speed at which they need to move. During the research, the scheduling of departures orchestrated the rhythms of passengers in the station and conducted the patterns of emergency-running. On the diurnal scale, the fluctuations of emergency-runner rhythms coincided with wider station and commuting patterns, themselves reflecting the wider social and cultural rhythms of work (Edensor, 2011). Perhaps unsurprisingly, this entailed that the morning and evening commuting peaks produced the highest prevalence of emergency-running, with an average of 36 runners per hour (RPH) and 44 RPH being witnessed respectively. The 12:00–14:00 period averaged 23 RPH. Such figures are slightly misleading, however, as the flow of emergency-runners was anything but steady. A RPH measure hides the impact of specific train departures on rhythms. This can result in huge differences in how much running occurs from one minute to the next. The fast service to London, which departed Platform Five four times an hour, was the largest stimulus in this regard with the peaks and troughs of emergency-running rhythms seemingly correlating to its departures. In essence, the temporal design of train departures creates the deadline which passenger must meet, and lays the foundations for the conditions of emergency-running.

Semiotics

Semiotics relates to the signage systems coded into material environments (Jensen, 2013). These can be physical signs, auditory signs or aspects of places interpreted as part of the signifying dimension. The semiotic system in the train station is a system of representation of train details 'read' by those engaging with it in order to make sense of the situation and interpret how they should act. Just as road signs within the Highway Code can help drivers understand the situation on the road and act accordingly, the diverse semiotic system within Guildford train station is designed to help passengers gauge how much time they have and hence how fast they need to move. This section will discuss the key signs prominent within the station and the ways in which their presences and absences influence passengers' abilities to judge their progression against their intended departure. A key point to note here, however, is that the sign system in the station is designed to be read at walking speed (3–5 mph), which influences, among other things, the size and ultimately perceptibility of the signs (Venturi *et al.*, 1972), an attribute with interesting results when participants need to move beyond walking pace.

The electronic information display boards were the most directly engaged with aspect of the semiotic system. The Ticket Hall displays the 'Next Fastest Train To' for all the destinations served by the station. This information is accompanied by a platform number, scheduled departure time and expected departure time. Throughout the rest of the station, the boards displayed the same information but organised by the chronologically forthcoming departures. These displays are the most accurate and direct way for passengers to check their progress and thus can be a source to encourage emergency-running when departures are imminent or a source to discourage and even halt emergency-running when time is plentiful:

A man in his late 20s wearing work attire and trainers jogs in, slowing to eye up the display. He takes his earphones out and scrutinises the board closer before popping them back in and sprinting, slaloming around two others before getting to the barrier.

(Ticket Hall 09/07/2014, 16.43)

An older woman in a body warmer, walking trousers and running trainers turns the corner into the station running and looking very red in the face. She glances at the board and continues to run. She suddenly stops, looks again at the display and proceeds to walk.

(Ticket Hall 09/07/2014, 07.54)

What is also noteworthy about these incidents is the need for both subjects to slow down in order to properly read the displays. The size of the text requires such slowed mobilities—it is designed to be read by the walking subject.

Interestingly, the displays did not impart information regarding the time required to reach particular platforms, as is increasingly common in signage at airports. This is something that would be useful within Guildford station, particularly as the actual distance needed to be covered in order to reach a platform from the Ticket Hall is far greater than the visible proximity suggests. The gap between visual and actual distance, coupled with the lack of information further encouraged emergency-running, particularly by those less familiar with the station layout. Despite there being a huge reduction in the number of passengers in the station during the midday hours, the RPH did not reduce by a corresponding rate. This is most likely because passengers departing the station at these times travel less frequently than the daily commuters observed in the morning and evening peaks and as a result they may be less familiar with the station layout. Coupled with the semiotic absence, the user thus has little certainty of what to expect, and it is this uncertainty that produces emergency-running.

Sounds and auditory clues are similarly part of the signage system at a train station, providing insight into the imminence of a train departure and a passenger's likelihood of making it. However, it is a more confusing signifying dimension to interpret. Not only are there a range of different auditory cues in the station, but there is an unequal geography to where the clues are present and absent. Speakers on individual platforms make three identical announcements for approaching trains. The first around 150 seconds before its arrival, the second around 20 seconds before arrival, and the third when the train is stationary on the platform. However, unless punctual enough to be present for all three announcements, passengers can be unaware of which announcement they are hearing and can be encouraged to run unnecessarily:

> The first announcement of the 0734 train causes a man and woman half way along the ramp and with perfect vision of the platform to begin to jog with concerned looks. Their heads and torsos immediately drop and lean forward as the uphill slope contorts their bodies. At the crest of the incline, both reduce back to a walk with the train not yet arrived.
>
> (Platform Five 14/07/2014, 07.30)

Others, however, most likely habitual travellers, seemed calm when these announcements were being made, often waiting for subsequent auditory signs of a conductor's whistle, signalling that the doors are about to close, or even the noise of beeping doors, signalling their imminent closure, before being persuaded to break into a full run:

> A woman with blonde hair and in a black suit holds her handbag in her left hand and a newspaper in her right. Despite the train already being on the platform, she confidently walks up three-quarters of the ramp. However a toot of the whistle and the beeping of the doors a few seconds later results in a frenzied head glance to the left, spotting the train and an even more frenzied run. She struggles to accelerate on the incline but on the flat she covers the five metres to the nearest open door more easily.
>
> (Platform Five 14/07/2014, 07.06)

Once again, absences in the semiotic system leading to uncertainty are productive of emergency-running. To take another example, all auditory clues to a train's imminence in the Underpass are muffled and become replaced with the echoing and reverberating sounds of footsteps, chatter and the deep, imposing rumble of suitcases. Not only does the noise of the rumbling suitcase become easily confused with that of train moving above, but it is also difficult to tell from the Underpass into which platform a train is moving. Combined with the lack of any visual clues, the confused auditory signals have a significant influence upon emergency-runners.

As these examples demonstrate, the semiotic system at Guildford train station is partial and at times confusing, leading to uncertainty, particularly in less habitual users. It is this uncertainty around the relationship between movement and time that transforms walking into running. Accordingly whilst the affordances of design are not universal, it is the relationship between different designed elements that produces differently mobile subjects.

Material site

As suggested by the location descriptions, the material site and physical layout of the train station contains within it many affordances for the possibility and impossibility of emergency-running. Three of the sites within the study (the Gateline, the Underpass and Platform Five) are linear spaces. These are design layouts aimed at channelling passenger flows between different areas of the station. They are efficient at dispersing passengers in a directed manner and lend themselves more readily to performing unhindered emergency-running if necessary. The Ticket Hall on the other hand, lacks such linear channelling and directed dispersal. Rather it encourages (and, in cases, requires) the station users to engage in a range of different activities which demand different forms of mobility and immobility, such as standing, queueing, chatting, shopping, etc. The promotion of such activities, and the presence of others performing these activities, can hinder emergency-running, evident in the performance of a range of walk–run–stop–run–walk sequences as people navigated the site and requirements of the space.

The design of material objects within the sites also influence the inclination or possibility of emergency-running. The most obvious example of this is the ticket barriers at the Gateline. Despite its simple (in principle) operation as described earlier, negotiating the ticket barrier is a task fraught with human and object deficiencies. The short opening time of the gates (aimed at stopping multiple people entering on one ticket check) means that bags and other accoutrements often get caught when they are closing. Having a ticket rejected by the barrier is similarly a common problem; also the direction a gate is serving (entry or exit) can be misread: green arrows or red crosses indicate the direction of travel yet these are too small to be read at running speed, suggesting an object designed with a walking user in mind. Such events can inflict further stillness onto a moving body; something intensified when already moving at speed:

> Three teenage friends run to the barrier shouting 'rush rush rush'. One gets through and begins to sprint but halts when the others' tickets get stuck. The girl doesn't wait long and sprints off towards the underpass as the others shout 'wait for us'. After help from the attendant, their tickets are finally sorted and the other two sprint off.
>
> (Gateline 24/06/2014, 16.31)

Such is the unpredictability of negotiating the barriers at speed that the sight of people only breaking into a run once it had been passed was extremely common. In many ways it marked a threshold in navigating the station; discouraging emergency-running before it and encouraging it afterwards.

The topography of the material sites also proved a significant affordance in encouraging/forcing as well as discouraging emergency-running. The majority of the station is flat, a topography relatively neutral when considering its impact upon running. However, entering and exiting the Underpass entails encountering both downhill and uphill slopes, a design feature with a major impact upon the experience of emergency-running. Uphill ramps warp bodies grappling its gradient and downslopes increase the pressure exerted on knees, exaggerating balance-reflexes or bodily contours:

> A larger man in his fifties wearing a cream suit and fedora hat sees the train on the platform when half way up the ramp. Through his sunglasses I see his eyes pop wide and he begins to run. Gaining momentum against an uphill gradient proves tough as his legs struggle to get much height.
>
> (Platform Five 14/07/2014, 13.33)

> A larger woman in her 20s wearing a short blue dress, heels and a trench coat runs awkwardly downhill. She is practically tip-toeing with short steps and an arm that swings in and out like a hinge to counterbalance the heavy bag on her right arm.
>
> (Underpass 11/07/2014, 08.28)

Those entering the Underpass must travel downslope at a gradient at which gravity seems to have an impact on the speed at which one travels. This leads to many people speeding-up downhill, making it easier for those already running, nudging those who are contemplating it and involuntarily forcing some walkers to break into a jog. Contrastingly, an uphill slope causes a slowing effect on bodies. This is both physical, climbing a hill is physically demanding, difficult and tiresome; and mental, the prospect of running up a hill is unappealing to many. This resulted in a marked reduction in the numbers of emergency-runners wanting to, or capable of, running on uphill portions of the station:

> Prompted by some passing teenagers, a woman in her fifties attempts to run laden with bags that her body fights to contend with. She slumps, her legs low as she struggles uphill. She walks for a few steps, readjusting the bags she continues her run, walking once more as her bodily rhythms make her bags fall once again.
>
> (Platform Five 14/07/2014 18.00)

As demonstrated, the material design of spaces and objects in Guildford station both encourage and discourage emergency-running in various ways. Yet, design is not simply autonomous, dictating when running should occur and when it shouldn't; evident in all of these vignettes is the importance of hybrid mobile bodies in engaging with design. As these accounts attest, users move through the station with varied accoutrements related to the art and craft of train travel, from clothes, to bags, to bikes and so on (Watts 2008). These are often functionally and aesthetically refined objects, but not necessarily special-ised to make running desirable, easy or efficient. Accordingly, each part of the emergency-runner assemblage interacts to impact a person's ability to run if required and thus is a constitutive part in the production of emergency-running in a station.

Conclusion: the design of certainty

In rethinking the contemporary role of running, this chapter has taken as its chief concern the production of running as an everyday emergency form of mobility in a train station. It has sought to interrogate the very mundane details about how mobilities actually happen by paying close attention to the affordances of mobili-ties design in the taking place of the spatially discordant act of running in a train station. It reveals both that the train station is an extraordinary site of running mobilities (though not necessarily running as we know it), and that running is an integral aspect of rail mobilities, though paradoxically one that is discursively discouraged whilst being materially produced. In exploring the production of emergency-running in a station, the influence of three aspects of mobilities design have been examined. In their presences and absences, the temporal staging, semiotics, and material site of the train station intersect to produce certainty and uncertainty. Running in a station only occurs due to a lack of certitude that peo-ple have that they can catch their intended train without increasing the speed of their mobility. Much uncertainty emanates from the unpredictability of navigating the station itself and negotiating the various objects, obligations and distractions that it presents, rather than due to pure unpunctuality. Of the over one thousand instances of emergency-running observed in this study, very few required running to be used for the entire journey through the station in order to make a depart-ing train. Much more commonly, emergency-running either only began once the most unpredictable areas of the station had been passed (such as the Ticket Hall and Gateline), or it stopped once certainty of one's fate had been assured. Emergency-runners, thus, were highly attuned or habitualised to the affordances that mobilities design offered in gauging certainty and uncertainty.

Certainty and uncertainty are both produced in the entangled presences and absences of the three aspects of mobility design discussed above. The temporal staging can produce certainty of a train's departure when everything is running smoothly, yet uncertainty when delays occur. In various ways the semiotic sys-tem at the station enables passengers to gauge their progress against their intended train, most commonly through use of digital displays. Yet the inconsistency and incompleteness of these signs, along with the clarity of them (particular auditory ones) can have the opposite effect, producing uncertainty through an inability to judge the relationship between mobility and time available. Finally, the varied material sites throughout the station produce both spaces which are simpler and more predictable to navigate, and those which are more difficult and unpredictable due again to the fact that it is not always possible to predict how long it will take to navigate such spaces.

Mobility is produced in the interaction between people and design; it is, after all, an improvisation and an accomplishment. In appropriating a space designed for walking in a hybrid assemblage designed for rail travel, emergency-runners are improvising in an encounter with design. Running in a train station requires the interpretation of the semiotic cues and practical affordances of a site in a judgement of whether running is required or not. Exploring this everyday form of emergency mobility reveals much about the ways in which mobilities actually occur and the impact that design has in producing these mobilities.

References

Adey, P. 2016. Emergency mobilities. *Mobilities,* 11: 32–48.

Bale, J. 1994. *Landscapes of modern sport.* Leicester: Leicester University Press.

Bale, J. 2003. *Sports geography.* London: Routledge.

Bissell, D. 2007. Animating suspension: waiting for mobilities. *Mobilities,* 2: 277–298.

Borgers, J., Vanreusel B., Vos, S., Forsberg, P. and Scheerder, J. 2016. Do light sport facilities foster sports participation? A case study on the use of bark running tracks. *International Journal of Sport Policy and Politics,* 8: 287–304.

Carléna, U. and Maivorsdotter, N. 2016. Exploring the role of digital tools in running: the meaning-making of user-generated data in a social networking site. *Qualitative Research in Sport, Exercise and Health.* Online First. DOI: 10.1080/2159676X.2016.1180636.

Cidell, J. 2014. Running road races as transgressive event mobilities. *Social & Cultural Geography,* 15: 571–583.

Cook, S. 2016. Run-commuting. In *Minicars, maglevs, and mopeds: modern modes of transportation around the world.* Ed. Sultana, S. and Weber, J. Santa Barbara, CA, ABC-CLIO/Greenwood Press, pp. 255–258.

Cook, S., Shaw, J. and Simpson, P. 2015. Jography: exploring meanings, experiences and spatialities of recreational road-running. *Mobilities.* Online First. DOI: 10.1080/17450101.2015.1034455.

Cook, S., Shaw, J. and Simpson, P. 2016. Running order: urban public space, everyday citizenship and sporting subjectivities. In *Critical geographies of sport: space, power and sport in global perspective.* Ed. Koch, N. London: Routledge, pp. 157–172.

Edensor, T. 2011. Commuter: mobility, rhythm and commuting. In *Geographies of mobilities: practices, spaces, subjects.* Ed. Cresswell, T. and Merriman, P. Farnham: Ashgate, pp. 189–204.

Jensen, O.B. 2013. *Staging mobilities.* Abingdon: Routledge.

Jensen, O.B. 2014. *Designing mobilities.* Aalborg: Aalborg University Press.

Latham, A. 2015. The history of a habit: jogging as a palliative to sedentariness in 1960s America. *Cultural Geographies,* 22: 103–126.

Office of Rail and Road 2016. *Station usage 2014–2015 data* [online]. Office of Rail and Road. Available from: http://orr.gov.uk/__data/assets/excel_doc/0019/20179/Estimates-of-Station-Usage-in-2014-15.xlsx [accessed 23 June 2016].

Qviström, M. 2013. Landscapes with a heartbeat: tracing a portable landscape for jogging in Sweden (1958–1971). *Environment and Planning A,* 45: 312–328.

Qviström, M. 2016. The nature of running: On embedded landscape ideals in leisure planning. *Urban Forestry & Urban Green Planning,* 17: 202–210.

Sport England 2012. *Satisfaction with the quality of sporting experience survey (SQSE 4) Results for athletics: trends 2009–2012* [online]. Sport England. Available from: http://www.sportengland.org/media/110940/athletics.pdf [accessed 27 March 2014].

Trott, K. 2015. *2011 Census in Surrey – first results* [online]. Surrey-i. Available from: https://www.surreyi.gov.uk/ViewPage1.aspx?C=resource&ResourceID=928&cookie Check=true&JScript=1 [accessed 3 May 2016].

Venturi, R., Brown, D.S. and Izenour, S. 1972. *Learning from Las Vegas*. Cambridge, MA: MIT Press.

Watts, L. 2008. The art and craft of train travel. *Social and Cultural Geography*, 9: 711–726.

6 Design mobilities via 3D printing

Thomas Birtchnell, John Urry and Justin Westgate

Introduction

In this chapter we take on board the challenge to explore how mobility has "emerged as an object of knowledge", and how forms of mobility are given meaning through design knowledge materialised into physical objects (Cresswell 2006: 2). In order to do so, we examine debates around the mass manufacturing and spread of 3D printers; products that allow individual users to print their own objects with potential consequences for current concentrated systems of production. We argue that stakeholders give meaning to the potential dispersal of design embodied in the 3D printer through discourses of education, empowerment, and democratisation. The chief argument is then that the 3D printer as an artefact "mobilises design" because of its potential to enable design knowledge to travel, disperse and be customised, in this instance from something done by designers in studios to consumers at home or in print shops or spaces near at hand.

Mass-produced objects are heavily personalised and given authentic meaning by users. The sheer extent of massification and seeming inauthenticity of mass-produced and mass-mobilised products means that users of, and experimenters with, 3D printers are engaging in shaping a new industrial future. This emerging discourse is underpinned by decentralised production in the homes of, or in proximity to, consumers who become designers and producers in the process of instrumentalising the technologies. In this chapter we also point out the limits to this slightly utopian ideal with regard to the limits of existing technology in terms of materials, technical knowledge, object size, speed of printing, cost comparison and so on.

Advocates of "open" hardware and software, such as the inventors of the "BeamMaker" low-cost yet high-resolution photopolymer prototype 3D printer, attribute its potential for "democratisation" to bring "digital manufacturing to everyone" with the "advancement of society" achievable through "access to education and the means of production" (Calderon *et al.* 2014: 245). Evidently there are egalitarian principles behind the deployment of such language implying that the ubiquity of 3D printing could equate to the movement of power from the few to the many. In response we contest that it is necessary to also talk to consumers about their interest in engaging with the means of production in order for such power to move to them. Indeed, democratisation as a process of dispersal relies on a number of related things "moving", in this instance, both the production facility and design expertise. At present, however, whilst the production facilities (in the form of the 3D printer) have the potential to reconfigure consumers as designers, the knowledges required do not appear to be moving as far or fast as 3D printers themselves. Hence we argue that democratisation of design/production remains uneven.

In studying this topic we are sensitised to charges of "technological determinism" and respond that "technology and its uses are socially constructed, have different effects across space and time, and social problems cannot simply be solved by technology but also need social and political solutions" (Castree *et al.* 2013: 505). In this instance, 3D printing can be viewed as a socially constructed panacea for all sorts of issues around identity politics and materialities (alienation from the design process and brand culture), inequalities in production processes (chiefly in low-cost and labour-rich Asia), sustainability and circular economies, and individual engagement with learning to "make" objects drawing on notions of craft expertise. Thus on the one hand 3D printing is being mobilised as a technology for the people, but at the same time is constructed as an "individualising" technology enabling authenticity to be created for the user.

The research that informs this chapter was a "vox pop" exercise (see Gibson *et al.* 2012) combining informal semi-structured interviews with the wealth of textual data (pamphlets, posters, flyers, banners) available at an industry tradeshow—the 3D Print Show—alongside analysis of related social and material culture: demonstration objects and marketing slogans. The tradeshow features in cities around the world—London, Paris, New York, Berlin, Dubai, Mexico and others—to showcase applications and developments in the technology. The tradeshow we attended ran over three days in September 2014 catering to stakeholders from a range of sectors and expertise. Held in a large display space in the heart of the City of London, data was collected at a research stall on the show floor alongside other 3D-printing exhibitors. Our research involved thirty-four interviews with participants at the show linked to 3D printing through either paid attendance as a visitor or exhibitor (Table 6.1). A quarter of interviewees were female and two thirds were in their mid-twenties to mid-thirties. Conversations were undertaken with a range of attendees across the three days to help source responses.

Table 6.1 Phase two—vox-pop interviewees

Sector	Number
Education	5
Medical	33
Engineers	5
Designers	4
2D printing	3
Fashion	3
3D-printing industry	5
Media and research	6
Total	34

Note: We recognise that these "categories" are not fixed: some participants operate across sectors (see Gibbs and O'Neill 2014, 104).

In recorded interviews ranging from 10 to 30 minutes, passers-by were invited to reflect on their experiences and perceptions of the exhibitions and the various materials made available to them alongside the various 3D-printing technologies and products on display. Themes during the interview were focused around key areas of interest: the subject's involvement with 3D-printing technology; their motivation for attending the expo; their reflections on exhibition offerings; impressions of the difference between different forms of 3D printing; thoughts on the benefits of 3D-printing technology to manufacturing and its wider potential social impacts. Questions catered to the individual's particular background, that is, they were adjusted to suit the operational area of each interviewed subject, allowing us to tease out more relevant details of their specific engagements with the technology.

The structure of the chapter is as follows: the following section introduces the reader to 3D printing and highlights its grounding in rapid prototyping services. A third section introduces the empirical case study material and highlights an emerging discourse around 3D printing as potentially shifting endemic power relations through the dispersal of production facilities. The fourth section considers where 3D printers are actually moving—namely, to design and engineering firms specialising in limited runs of boutique, unique or "custom" objects and form, fit and function prototype models—and examines the associated more limited movement of knowledge and expertise. The fifth and final section begins to draw conclusions around the potential of 3D printing to mobilise design, presenting a somewhat more tepid narrative of co-production suggesting that both design knowledges and artefacts remain limited in the extent of their dispersal.

The emergence of 3D printing

The focus of this chapter is 3D printing as a social science topic (see Gress and Kalafsky 2015). In the early 1990s, new desktop (or benchtop) manufacturing technologies were developed for businesses to produce accurate prototype models from computer design data in a matter of hours, in contrast to conventional prototyping methods requiring weeks. Put simply, designers would divide a 3D computer-aided design (CAD) model of an object into thin, printable horizontal cross-sections. The stacks of wafer-thin laminations were then fused with a laser to create a 3D object (Miller 1991). The physical prototype allowed designers to see and hold physical models of part designs making "fiction fact fast" as one industry commentator put it (Wohlers 1991).

With the arrival of FDM[1] desktop units in consumer markets, 3D printers able to produce plastic objects in near-saleable quality became mainstream products. "If you can imagine it you can make it" as the slogan of a popular company (Makerbot, recently acquired by Stratasys) would have it. There are also industrial uses for DMLS and EBM printers able to produce single instances of objects with high resolution of detail and as critical parts in metal, for instance in aerospace and automotive applications. 3D printers of these kinds are an applied technology for custom parts in high-end objects including passenger aeroplanes, luxury automobiles and extra-terrestrial vehicles (e.g., NASA's "Curiosity" Mars Rover). 3D printing also has many medical applications both in copies of the body and in replacement parts (Lupton 2015). Companies such as Shapeways using industrial 3D printers already cater to consumers wanting single or limited numbers of objects in some cases with geometrically complex designs or exotic aesthetics.

Industrial engineers, architects, designers and other users requiring models for testing and bringing concept-to-production use 3D printers in their everyday work. Also 3D printers feature in forecasts of a pending industrial revolution as a disruptive force in systems of production, distribution and consumption. One example of this is in the brainchild of Baback Elmieh, an ex-Google employee, titled "Nascent Objects". This is a design platform involving a dozen "modules" representing the basic components of a great many objects: sensors, computer processors, microphones. Surveying 600 objects on the market in 2012, Elmieh's team found 80 per cent could be made from just 15 common electronic components. Using 3D-printed brackets the platform allows the user to drag-and-drop modules into their own object's skeleton and then the software calculates the circuitry pathways, which are printed into a chassis. This idea represents the automation of much knowledge work currently undertaken by professional designers. As the example of Nascent Objects demonstrates, there are now design innovations that have implications for many other aspects of the global production–distribution–consumption triad (Stinson 2016).

Potentials of 3D printing: moving production

Our data reveals that participants conceived the benefit of 3D printing in terms of the movement of the means of production from concentrated industrial contexts into dispersed design contexts, ultimately enabling egalitarian ideals of democratisation—envisaged as a process of dispersing power from the few to the many. They discussed 3D printing as enabling democratisation in three main ways: dispersing the means of production; dispersing design knowledge; and improving quality. In this context, 3D printing is something different to other manufacturing technologies (and indeed, as participants noted, is not cost comparative to the bulk volume manufacturing of many identical items) by returning the possibility of production to consumers in their homes or at least near to hand. The interviewees themselves identified this as a social "movement":

> I think this is a very important movement, I think the democratisation of technology which used to only be available if you're an engineer within an aerospace company, super high-end, very locked up and now it's making its way out into the world.

By moving the factory from centralised facilities, democratisation here is in tune with the notion of being "off the grid" of global production, distribution and consumption systems. Being off grid does not necessarily indicate a lack of interconnectedness with the rest of society; rather it constitutes a sense of independence in re-assembling everyday life (Vannini and Taggart 2015). A key factor in achieving this independence through 3D printing is the recent emergence of open-source machines (notably the Reprap and the many start-ups that used this as the base of their kits) due to the expiry of patents in FDM processes.

Some interviewees were more cautious about this idea of democratisation of production through 3D printing, drawing parallels with the computer revolution and the complexities of profit and not-for-profit interests in the current system. As one interviewee reflected, open-source technologies do not always beget democratisation. The software market continues to be dominated by monopolistic corporate entities even though there have been open-source alternatives available for some time:

> Yeah, but like the thing about the open-source movement, in software, it was like a lot of talk but there's still Microsoft and it hasn't changed. It has changed some, but not so much that they believed before.

Another claim made in relation to the movement of production away from centralisation relates to issues around labour standards and the expertise (or lack of) involved in the current bulk volume production system. Concerns over the exploitation of labour in mass manufacturing has led to the production of a new class of objects, known colloquially as "fair trade" products, whose quality derives from the guarantees given that basic labour conditions are met in centralised factories or in some cases cottage industries. The idea that 3D printing as a manufacturing technology may reduce reliance upon unregulated poor quality labour on assembly lines has attracted some scrutiny (Bogue 2014). The sheer complexity of mass manufacturing caused some in our interviews to question the scope of 3D printing to ever be able to make quality objects in a decentralised fashion:

> Oh, 3D, we want to do. I think we … for the moment the technology is okay on extruding plastics and on polymers, but not on metals, not on textile and all that kind of thing.

As this suggests, 3D printing makes little sense as a replacement for the production of large numbers of identical items and will not be able to make the majority of the objects appearing in people's everyday lives unless key technical challenges are overcome. Others were more upbeat, noting that computer-controlled welding in the home offers a similar functionality to SLS 3D printing in metal:

> I think what's coming, the next state will be laser SLS, what they're doing with the metal they can do that with plastic and all sorts of things, when the price of those machines become accessible to home users that's going to blow this out of the water.

As these interviewees suggest, the idea that 3D printing may enable better quality products to be produced and without exploiting labour is contested. Certainly most products made from 3D printing would fall into the category of not being "Made in China" or labelled as such, although ironically many 3D printers are made in bulk in Asia as consumer products: instead 3D printing relies to a large extent on the concentrated systems of production it potentially replaces.

Whilst contested, what we see emerging is a discourse of 3D printing as potentially democratising because of the dispersal it can facilitate. However, at present there is currently limited potential for design and manufacturing to move into the home and hands of the consumer due to factors such as complexity and only being applicable to working in certain materials. That being said, there are many possible avenues wherein democratisation does occur due to some degree of decentralisation (of both "facilities" and design knowledges) and a consequent engagement with 3D printing in the wider society as the means of production is mobilised. It is this we turn to in the next sections.

Movement of the "facility" and design expertise

Easy access to objects is a recent phenomenon, falling within the lifetime of most of our interviewees. Whilst participants discussed the potential of 3D printing to colonise new spaces, in this section we consider where 3D printers are actually moving to. Interviewees discuss in this section the potential of 3D printing to mobilise design, but also that it is currently limited because people are not designers and the technology does not straightforwardly help them become designers. The argument here is that there is a limit to how fast 3D printers will cause a "revolution" depending on how dispersed the 3D printer becomes, but also the expertise required to operate and create objects with it.

Once a generation had grown up with computing and it had entered schools, office and homes and become "normal", there was a consequence for the extent of digital technology in society. It was clear that for interviewees, as in the early days of computing, the decentralisation of the "means of production" was creating both anticipation and confusion. One potential emerging with 3D printing is the ability of almost anyone to access the tools of production. Here there is a sense of opportunity:

> Yeah, and it took a while before … people only had them in the office and then all of a sudden now you can have one in your house. It took a while but now we're at that point. I mean you could easily see this trend in the same way. They all make the comparisons, the computer or the car. Firstly they were all kits in the beginning.

Other interviewees, however, noted that 3D printing in the home accessible to the majority was still a long way off:

> I guess the point is that I see a lot of other machines, the 3D printers and the hype is that it's just going to be replicated in your house and we're decades away from that.

So whilst many felt that the 3D printer could move production into the hands of more if not the many, one of the barriers to doing so was the imaginative immobility related to 3D printing as a new technology:

> I think there's a bit of confusion in the public sphere. It's like people don't know quite what to make of it. It's so kind of—I know it's not new but it's new to the general public. So it's like they don't know how to process it, it's like the position of it in their minds is a bit vague, and they associate it with geeks and our ties or something like that.

This idea that the lay person did not know what to make of 3D printing also calls into question the assumption that 3D printing could democratise production through its educational potential, enabling everyone to become a designer and producer of products. The empirical material here points to the educational potential of 3D printing as a key driver of 3D printing in the moving of design into the home, though suggested more so by the non-professional individuals we interviewed. Here 3D printing was seen as an exciting and novel activity that young people can engage with in science, technology engineering and mathematics subjects (Rosen 2014). Beyond expectations of changes to how objects are moved was the theme that once children had experienced 3D printing, they would no longer see the need for containers and logistics for having objects "pre-fabricated":

> As an educational resource it's huge. My four-year-old loves the 3D printer. To him it's like the coolest thing ever. You can make stuff. So get in the matrix of that level and then imagine what they're going to do with that. Imagine growing up that small knowing that you can just make whatever you want within reason, obviously, within the capability.

This is not just about the output—producing novel objects— but also the practice of designing and customising objects as a learning process. One hobbyist spoke of his involvement at a community level helping to organise and run classes for young people that introduced them to 3D printing with the goal of developing horizontal skills such as programming and electronics. Other attendees we spoke with mentioned education as a factor, whether as an additional motivation for attending the expo or as a reason to bring children along—exposing them to an emerging technology with hopes of stimulating young minds.

Others were more sceptical of the idea that the presence of the printer alone would educate and enable a dispersal of design expertise. A key caveat here is the complexity involved in designing and producing objects, still out of reach of baseline consumers and users lacking expertise or training in design software. Another factor is the complexity of the machines, which—similar to paper printers—will jam, malfunction or respond to human error:

> Yeah, this is the big one, getting stability in your prints. You have to think: the non-technical people aren't going to be tinkering with every little function of it. They might get frustrated pretty quickly when they realise it's just not working.

Again this can be seen to mirror the evolution of other technical innovations—computers and the Internet—with proclamations of their educational potential which are rarely realised. Whilst objects such as 3D printers do configure their users as social intermediaries, required levels of expertise and creativity do not automatically move with objects.

Co-creation: between concentration and dispersal

Whilst many stakeholders saw a democratising potential to 3D printing, they also saw the limitations to these potentials being fulfilled, partly because of limited movement of 3D printers into the home, but more because it was felt the design expertise and creativity required to fully democratise production would not disperse in the same way as the printers themselves. What emerged from this was a compromised narrative of more limited dispersal into communities rather than households:

> [I]f you break something and it's discontinued you can just copy it or take it to a shop and have it copied which would be brilliant. I think more not about having one at home but going to a place where you can get something produced.

What we see here then is a more pragmatic claim about the relative dispersal of the 3D printer as production facility; not down to the level of the home as some claim, but rather a weaker form of dispersal where 3D-printing shops might open up in the high street. Our interviews showed that 3D printing was being taken up by small-scale market actors who saw niche application for customised products, jewellery or fashion accessories, and personalised functional objects: mobile phone earpieces, wrist braces. The entrepreneurs we spoke to evinced a range of innovative thinking as to how to marketise the customisability of 3D printing—though the common challenge was how to find and manage a sustainable business model given the often small-scale, boutique nature of their offering.

Certainly low-end printers have become affordable to the everyday person, but even the more advanced and expensive processes are available at cost-effective prices through 3D-printing services, developed initially from rapid prototyping (e.g., the company Shapeways). This access to production allows people to explore the possibilities of materialising ideas they have. More than just hobbyists playing around, within our interviews we encountered a number of individuals who fell into the "inventor" class—who were working on somewhat advanced projects with specific applications. One engineer, for example, talked about a transmission system he was working on for wind-powered vehicles. Such instances show that the technology does indeed open up the possibilities for more advanced and innovative production—even if this isn't done with the intent of marketisation. As another interviewee went on to state: "In the meantime this is a tool that a professional designer can really sink their teeth into and make awesome things with".

In line with this enduring focus on the expertise of the designer, we encountered examples where marketisation was a focus, but with the intent of a co-creative production. Here personal customisation is an added value to a service. The customer may not have an extensive range of creative skills, but a production service can provide the ability for certain parameters of an item to be fine-tuned by the customer. One company we encountered, for example, offered a service where customers could produce personalised vases by modifying a base model online to their requirements and then having it printed and shipped. This synergy between the ability to manipulate a virtual model and cost-effectively produce low-volume outputs was one we saw being taken up. To the customer, however, it was deemed a perfect fit, and certainly in an age of mass-production, viewed as an antidote with which designers are experimenting:

> So you don't need to be a creative person to have something that looks like ... you would just have to work with somebody, and we are these people who can work, I mean not work, but co-create something with our customers.

In the same fashion as the transition to telecommunications from face-to-face meetings in work—or word processing and paper printing—a great deal of experimentation occurs and a new technology rarely simply substitutes for another. As these accounts attest, many users are in the process of learning about how the new system can function in reality. The personal computer was initially a part of hobbyist cultures that only later created new commodity cultures—in gaming for instance. Personal computers were used for niche hobbies rather than for everyday necessities in which they now appear: communication, mobility, social networking, office-work and so on. The ability to use computers nowadays hardly requires computer-programming knowledge as it did in the past. 3D printing could also be following a similar trajectory from niche-level to become a dominant element in a different kind of socio-technical regime from the incumbent one. In all likelihood, people will continue to use both personal and mass manufacturing, just as they use both videoconferencing and face-to-face meetings or computers and printed text. Certainly at this stage, the evidence suggests that a complete transition and dispersal of production through 3D printing is unrealistic.

Conclusion

The crux of this chapter has been concerned with how movement (and 3D printing) is given meaning by actual users and those imagining themselves to be so in the future. With any emerging technology it can be difficult to actually imagine both the applications and implications of its use. Here the critical analysis of discourses of manufacturing and use allows an optic into how niche innovations impact upon society. The argument in this chapter shows we are dealing with different elements of a potential democratisation—suggesting a movement of power into the hands of many rather than few; however, what our empirical examples show is that whilst the "factories" can move into homes, they have limited capability and the expertise required to use them is still relatively concentrated in the hands of a few. Hence one of the key points our chapter illuminates overall is that democratisation requires more than the movement of the physical means, it requires the movement of knowledge and expertise as well. Of course, expertise is acquired through using the physical "factory" as objects configure their users so this may develop over time, but at present in the narratives of our interviewees there is still a separation of the two to some extent.

In our examination of the London 3D Print Show we were given insights about how stakeholders imagined the technology could have greater consequence than rapid prototyping. Certainly there were scores of small brightly coloured, single material, plastic objects on show: phone accessories, scaled models, toys and trinkets. What struck us from a more theoretical standpoint was how 3D printing allows us to manipulate matter in a way we have not been able to do so before. What 3D printing allows is for us to transfer ideas between a computational, virtual space and the material world. This gives us the ability to use more complex processes to compute and model objects and then give them form in almost any material. This opens up a range of possibility in what can be produced—we can create objects that, previously, we have not been able to produce, let alone imagine. In that sense, 3D printing both allows us and indeed invites us to move our imagination as to the types of things we are able to produce.

With regard to a broader picture of international trade, we do not assume that in a 3D-printing-driven transition in transnational trade, people will not still demand objects produced through assembly-line factory production and other techniques set apart from the consumer. Rather, 3D printing could disrupt the current system and make it unviable through drawing consumers away from the kinds of products it currently supplies to them *en masse* and cyclically (Petrick and Simpson 2013). At a wider scale, a systemic socio-technical transition would have knock-on effects in user understandings of the value of materials and energies (e.g., likely to be in the form of cartridges of powders) and how objects become recycled or simply waste for landfill once their lifespan is deemed over. At present it seems that 3D printing is adding to rather than reducing global freight because the objects being produced are in addition to rather than replacing existing mass-produced objects, and the trade in printers also requires freight movements. Systems are just not in place for 3D printing to be practised sustainably.

Note

1 For example fused deposition modelling (FDM), stereolithography apparatus (SLA), selective laser sintering (SLS), direct metal laser sintering (DMLS) and electron beam melting (EBM).

References

Bogue, R. 2014. What future for humans in assembly? *Assembly Automation* 34: 305–309.
Calderon, A., Griffin, J. and Zagal, J. C. 2014. BeamMaker: an open hardware high-resolution digital fabricator for the masses. *Rapid Prototyping Journal* 20: 245–255.
Castree, N., Rogers, A. and Kitchin, R. 2013. *A dictionary of human geography*. Oxford: OUP.

Cresswell, T. 2006. *On the move: mobility in the modern western world.* New York: Routledge.

Gibbs, D. and O'Neill, K. 2014. Rethinking sociotechnical transitions and green entrepreneurship: the potential for transformative change in the green building sector. *Environment and Planning A* 46: 1088–1107.

Gibson, C., Brennan-Horley, C., Laurenson, B., Riggs, N., Warren, N., Gallan, B. and Brown, H. 2012. Cool places, creative places? Community perceptions of cultural vitality in the suburbs. *International Journal of Cultural Studies* 15: 287–302.

Gress, D. R. and Kalafsky, R. V. 2015. Geographies of production in 3D: Theoretical and research implications stemming from additive manufacturing. *Geoforum* 60: 43–52.

Lupton, D. 2015. Fabricated data bodies: Reflections on 3D printed digital body objects in medical and health domains. *Social Theory & Health* 13: 99–115.

Miller, B. 1991. Rapid prototyping. *Plastics World* 49: 44–47.

Petrick, I. J. and Simpson. T. W. 2013. 3D printing disrupts manufacturing. *Research Technology Management* 56: 12–16.

Rosen, D. 2014. Design for additive manufacturing: past, present, and future directions. *Journal of Mechanical Design* 136(9): 090301 (2 pages).

Stinson, L. 2016. Nascent objects: an ambitious plan to stop us from wasting our gadgets. *Wired.* http://www.wired.com/2016/01/nascent-objects-an-ambitious- plan-to-stop-us-from-wasting-our-gadgets/ - slide-6 [accessed 13 January].

Vannini, P. and Taggart, J. 2015. *Off the grid: re-assembling domestic life.* New York: Routledge.

Wohlers, T. 1991. Make fiction fact fast. *Manufacturing Engineer* 106: 44–49.

Part II

Mobilising design

The mobility of design knowledge and practice

7 Why ship air?

Packaging design, mobilities and the materiality of void fillers

Craig Martin

Introduction

The constituent parts appear to be some sort of puzzle: three bits of wood with pre-drilled holes; a circular wooden disc; two more wooden discs, one with a single hole, the other with four; a wooden turn screw; a tube of glue; various cardboard packaging inserts; and protective materials. In separation they take on an almost abstract quality; sculptural suggestions. It's not clear what they are for, or how they might fit together. The instruction manual reveals more: an inexpensive, flat-packed IKEA Svenerik stool (Figure 7.1). Whilst accounts of IKEA have primarily stressed the debt to high-modernist design classics; the universal language of their visual instruction manuals; or the changing role of the consumer with self-build furniture, for this chapter my interest in IKEA products resides most significantly in the relationship between designed products, transportation and mobility. In particular, the development of their first flat-pack design in 1956 (Kristoffersson 2014, 43) and the subsequent proliferation of this design solution is rooted in the importance of transportation and distribution to the development of products themselves.

Efficient arrangement of the Svenerik stool's constituent parts offers IKEA economic advantage in reducing shipping costs. Rather than 'shipping air' the flat-pack configuration offers greater spatial efficiency. Marcus Engman, Design Manager at IKEA, argues that the transportation of their product ranges is absolutely central to the design of the products themselves (cited in Dezeen 2015): indeed as part of the product development process the designs of the products are often altered in order to reduce the size of the shipment package. Doing so can increase the number of individual products that can be stacked on a pallet, thus reducing overall transport costs. Engman states that reducing the size of product components by "one centimetre there could maybe mean [saving] 10 euros in the end on each and every product" (cited in Dezeen 2015). Whilst there is of course a marketing rhetoric at work, such statements demonstrate the inherent union between the development of consumer products, the mobilisation of products as part of the shipping process, and—crucially for my arguments—the critical importance of packaging design. One of IKEA's key innovations then has been the centrality of transportation to the design process. Logistics leads design. Why ship air when the spatial efficiency of flatness makes economic sense?

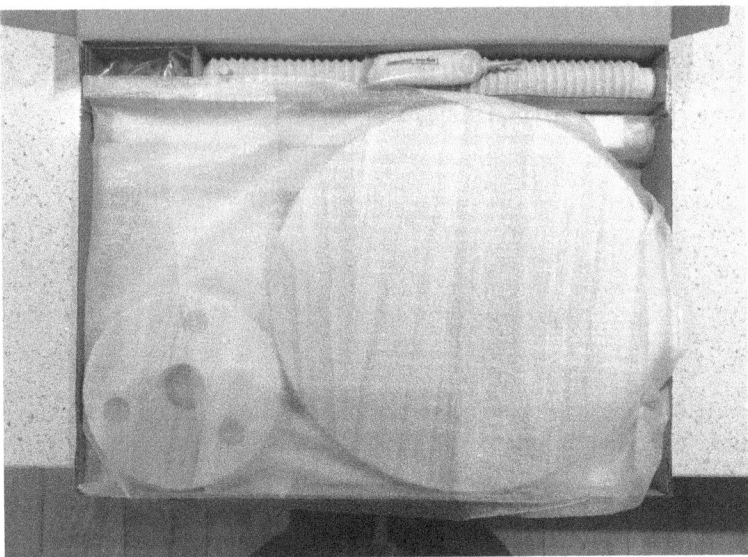

Figure 7.1 Flat-pack IKEA Svenerik stool.

The relationship between design, mobility and packaging evinces a range of crucial points about the role of design in mobilities studies. Fundamentally, there is a complex entanglement between how mobility is designed, and equally how design is mobilised. Based on this, my aims are two-fold. First, the chapter considers the centrality of packaging design to the mobility of designed goods, particularly their distribution and supply-chain mobilities. This stands in distinction to the traditional positioning of packaging design as a matter of taste, retail psychology and consumption (van den Berg-Weitzel and van de Laar 2001; Porter 1999). My approach to packaging protection is multi-scalar, echoing the differing scales at which packaging operates. García-Arca and Prado-Prado (2008, 375) identify three stages of packaging, with the direct protection of goods representing the first stage, namely 'primary packaging' (much like the IKEA stool). Moving up the scale, 'secondary packaging' deals with the grouping of primary packaging boxes into larger scale boxes, and the 'tertiary stage' involves the transportation of palletised loads, for example in shipping containers or other load units. Decisively, with the latter stage in particular there is a distinction between the spatial efficiencies of packaging design integrated into the product development process and the shipment of these within the wider freight supply chain. Above all, the chapter argues that greater consideration be given to the place of protective packaging within the study of design cultures and mobilities.

A second aim of the chapter is to investigate the importance of a crucial, but under-considered, aspect of packaging design: the materiality of packaging inserts themselves. Such mundane, seemingly throwaway artefacts are central to the mobility of designed objects. Although the example of the IKEA stool shows efficiencies in the spatial arrangement of constituent elements, the nature of these individual elements is such that there will inevitably be void spaces in the package that remain unfilled. This is where packaging materials are employed: to fill gaps and thus protect goods whilst in transit. The chapter makes an important distinction between 'void shaping' and 'void filling'.

The chapter is structured as follows. It begins with an outline of the importance of distribution and mobility—by way of circulation—to wider discussions of design and commodity culture. The second main section considers the importance of packaging and protection to corporeal mobilities, whilst also drawing out a direct relationship with the role of protective packaging design in logistics and supply chain mobilities. The final two sections build on my key points above about the distinction between a bespoke *shaping* of void spaces, and an ad hoc or improvised *filling* of void spaces.

Distributing design

The relationship between design and mobility is wide-ranging and complex. There are numerous moments of emergence where debates surrounding design (as both process and product) enmesh with mobilities. Perhaps the most clear-cut example is that of automobility, where the automobile stands as the paradigmatic manufactured object of corporeal mobilities (Urry 2004). Design is central to automobility: from the development of the car itself as an aesthetic form, engineering design and tooling for manufacture, marketing and promotional literature, through to attendant infrastructural design. Although not expressly named as such, Urry's (2004) rendering of the relationship between manufacture and infrastructure highlights an area of mobilities studies that is at present under-represented: the shipping and distribution of raw materials, components, and finished products (cf. Birtchnell *et al.* 2015).

Within the wider context of social geography, a number of scholars have addressed interrelations between production, the circulation or distribution of commodities, and their consumption. This literature—broadly identified as commodity chains geography (Hughes and Reimer 2004)—has sought to address *interconnections* between production, distribution and consumption. Whilst there have been significant discussions of the relationship between various nodes in the geography of commodity capitalism (see Hudson 2005; Hughes and Reimer 2004; Leslie and Reimer 1999), the specific literature on distribution has remained limited in scope, particularly from a cultural theory perspective. Despite Leslie and Reimer's (1999, 402) call for "the analysis of different sites, including production, distribution, retailing, design, advertising, marketing and final consumption", analysis of some sites remains undeveloped. As this chapter suggests, the interplay between design and packaging is crucial. It also underscores the important question as to where the activity of design itself resides within commodity chains. I argue that design pervades all aspects of the chain, including the design of 'mundane' packaging inserts.

The lack of an explicitly cultural focus within discussions of commodity chains has been redressed to a certain extent through the commodity *circuits* approach (see Crang 1996; Jackson 1999). In their study of the Sony Walkman, du Gay *et al.* argued that the "circuit of culture" (1997, 3–4), was constituted by five processes: representation; identity; production; consumption; regulation. In this literature there is a more expansive appreciation of the numerous sites and practices where commodity culture emerges, including advertising, the domestic realm and the media, but to this I would add the need to recognise the perceivably mundane materialities of cardboard packaging inserts. In so doing we will see the bond between sites of circulation and the place of packaging design. By delving more deeply into the materiality of packaging itself, we can move beyond the surface of commodities and consider the mundane material 'infrastructures' that propel them.

Packaging mobilities

The link between mobility and design is also central to Virilio's (2006) argument that the violent nature of corporeal mobility has been masked by developments in the design and engineering of transportation. He terms this "the corporeal 'packaging' of the passenger" (Virilio 2006, 54–55). If the 'packaging' of the passenger has been fundamental to corporeal mobilities, an attendant form of protective packaging also has been pivotal to the mobility of objects, particularly goods and commodities. Just as the upholsterer's art led to increasing physiological comfort (see Grier 1988) the packaging of consumer goods is a significant aspect of contemporary global capitalism. Most significant of all is the development of the intermodal shipping container—the centrepiece of international logistics (Martin 2016). However, for smaller-scale mobilities, packaging plays a fundamental role. Whilst historically, packaging design has often been at the forefront of technical and material innovations such as barrel design (Twede 2005), in the contemporary milieu, packaging design is typified by an aesthetically determined approach, concerned primarily with the outward signification of packaged products and their influence on consumer choice (van den Berg-Weitzel, and van de Laar 2001). An emphasis on the aesthetic coding of packaging design focuses all too readily on product packaging in the context of retail and consumer environments, overlooking the development of packaging material technologies.

By addressing the protective dimension of packaging, as opposed to symbolic signification, attention is shifted from consumption to distribution. Returning to García-Arca and Prado-Prado's (2008) three stages of packaging, a key determinant of the mobilities of goods is the need for protection during transportation. Lancioni and Chandran note that "packaging in international logistics can be defined as a 'unitisation system which begins with the shipper and ends with the customer'" (1990, 41). Such a description is denuded of the lived realities of how goods move through supply chains; and Lancioni and Chandran (1990) define a variety of 'interfaces' where packaging plays a vital role in international logistics. These include the initial manufacturing context where finished goods are moved to packaging locations in a factory; the intermediate loading of packaged goods onto pallets; the transportation of goods in shipping containers or other intermodal forms of shipping; and unpacking at the final point of destination. Emblem and Emblem (2000, 11) highlight a similar narrative, in which one of the fundamental aspects of packaging design is the need for goods to survive "storage and handling stages en route to the point of sale". This point accentuates the centrality of distributive space to contemporary consumer capitalism. It also emphasises how this aspect of the relationship between packaging and mobility is premised on warehouses, stockrooms, forklift trucks, shipping containers and vehicles, all logistical interfaces in the "international logistics package flow" (Lancioni and Chandran 1990, 41). Protection is central. Admittedly less alluring than the clichéd notion of design as the creation of spectacular aesthetic forms, the design of protective product packaging forms a central facet of logistics and supply chain management, but also is central to the wider notion of design culture and design thinking to which this chapter speaks. Although one might consider design thinking as primarily premised on specific problem-solving approaches to larger-scale products and services (Kimbell 2011), the same methods of design thinking and iteration are central to the design and engineering of packaging itself. Consideration needs to be given to the nature of goods to be protected, and this is where interrelationships between logistics and product development become significant, notably in the role of packaging design and engineering. Whilst the integration of protective packaging design and the development of products and goods is far from common practice, the economic and sustainability benefits of integrating them has been recognised by a number of global brands, including IKEA (Klevås 2005) and Volvo (Pålsson *et al.* 2013).

The field of package engineering itself is a scientific practice dealing with the design and engineering of materials for the protection of goods during shipment and storage (see Hanlon *et al.* 1998). As Ashley (1992, 67) notes, the growth of package engineering developed from the recognition of packaging as a constituent aspect of product development—so much so that in some cases packaging engineers work directly with product designers "to optimize the product/package combination and balance the cost of additional packaging and shipping fees against the cost of strengthening the product to withstand shipping forces". The symbiotic relationship between product development and packaging design is evident here, and Bramklev's (2009) use of the 'product-package system' term helpfully locates this. Although products will inevitably require some form of secondary packaging at certain points during storage or distribution, Bramklev's (2009) argument is that the link between the design and development of a product and its packaging as part of distribution should be integrated. Doing so provides a more efficient system where logistics (and by definition mobility) is part of the product development process. However, potential benefits obtained by integrating packaging design into product development are often not fully realised. For Bramklev, "when a package is needed, considerations are seldom devoted to it during the actual development of the product" (2009, 172). According to Klevås (2005, 124) packaging should be a central facet of the product development process, echoing

Ashley's (1992, 67) assertion that this reduces shipping and packaging costs as well as improving product-development lead times. Using IKEA as a case study, Klevås (2005) demonstrates how the company has been at the forefront of integrating packaging design into the product development process since the 1960s. Practicality was central to the development of the first IKEA flat-pack table—the Lövet—in 1956 when the table's designer could not fit the prototype into his car (Kristoffersson 2014, 43). To reconsider Engman's comments that "one of the biggest costs when it comes to producing or doing product development is logistics" (cited in Dezeen 2015) it is clear that the spatial efficiency of flat-pack furniture allows cost reductions in transportation and thus ultimately to the consumer. However, it is not solely in the context of flat-pack furniture that one sees the relationship between design and logistical mobilities in IKEA's operations.

Shaping the void: packaging materialities

Although the logistical efficiencies of the "whole package" (Klevås 2005, 116) are critical to the mobility of goods, such discussions have understandably taken a system-based approach (Saghir 2004), where the focus has been on systems to improve distribution and transportation. In these final sections I take a different tack and turn to the materiality of packaging materials themselves. By addressing packaging materials directly I argue that they form a significant part of the relationship between design and mobility, both in terms of the debates in the previous section, and as forms of mundane design in their own right. Whilst the discussions above have dealt primarily with the role of packaging design as part of the product development process and the logistics supply chain, here I directly address the nature of the packaging inserts. Crucially, by inverting the logic of design culture by addressing the seemingly mundane, throwaway packaging inserts we can further appreciate the complex mechanisms that pervade the mobility of designed goods. Although the symbiotic relationships between all aspects of the 'whole package' are critical there is something incredibly strange and disarming about the nature of de-contextualised packaging materials and inserts. By removing them from the context of the goods they are designed to protect they take on a singularity, a specificity of their own. By addressing packaging inserts as 'void shapers' we can begin to appreciate the intellectual significance of the space surrounding products as well as the products themselves. Echoing Vidler's (2001, 13) comments on the value of the void as a space worthy of study, my interest in packaging materials stems from the inversion of object relations that the packaging materials themselves produce. Within the study of design very little has been made of the physicality and materiality of packaging.

By focusing upon packaging, the space surrounding the object becomes just as significant as the object itself. In this context, Heidegger's (1971) essay 'The Thing' is a valuable tool for questioning how 'things' come to be. Using the example of a jug, Heidegger suggests that the object's task as a thing is to contain: "the jug's thingness resides in its being *qua* vessel. We become aware of the vessel's holding nature when we fill the jug" (1971, 166). Rather than the material form of the jug itself, the sides and the base, it is the emptiness or the void of the vessel that performs the containment (see also Knappett *et al.*, 2010). This may appear counterintuitive, for one might assume that in forming the jug the potter crafts the sides, the base and the handle. But Heidegger asserts that:

> No—he [sic] shapes the void. For it, in it, and out of it, he forms the clay into the form. From start to finish the potter takes hold of the impalpable void and brings it forth as the container in the shape of a containing vessel. The jug's void determines all the handling in the process of making the vessel. The vessel's thingness does not lie at all in the material of which it consists, but in the void that holds.
>
> (1971, 166)

Although a Heideggerian approach to package engineering may appear to be at odds with existing discourses on packaging logistics, his argument offers a valuable sense of packaging design's importance in understanding the wider notion of thingness and object mobilities. Just as the potter is effectively crafting the jug as void, in the field of package design and engineering there is a concomitant shaping of the void through the focus on the negative space surrounding the package. To understand the thingness of the product we need to look at how the void space is crafted. By considering the void spaces of packaging we can garner a greater sense of the relationship between designed products, their packaged state, and—most critically—how they move.

To begin, we can return to the wider contexts of package engineering and the 'product-package system' and consider the design of the packaging in relation to a commonplace electronic consumer item: a digital camera (Figure 7.2). Moulded paper pulp packaging is usually either brown or grey in colour with one smoother interior side and a rougher exterior. Formed from recycled craft paper or newsprint, moulded packaging offers strong protection for consumer goods such as electronics, being able to withstand drops of over 1 metre, and is particularly useful for precision applications (Thompson 2012, 22).

Beyond material tolerances, these seemingly throwaway pieces of paper pulp are of critical importance to object mobilities, and have power as things worthy of consideration in their own right. In photographing the packaging in such a manner their standing as distinct objects of study is emphasised. They are more than simply supports for the products. They have an intrinsic formal worth. So symbolic coding might form a typical reading of packaging design, alongside their distributive capacities these pieces of package engineering are important pieces of standalone design.

Closer inspection of Figure 7.2 shows the internal contours of the moulded packaging. Its form is determined by the relationship with the object it is designed to protect: it is the void-object. As García-Arca and Prado-Prado (2008, 375) note, primary packaging is designed to protect the product often through direct contact with the product itself. The camera resides in almost suspended animation; its lens section nestles tightly in the hourglass space in the image. Whilst the product is the primary focus, by investigating the void surrounding the product we can

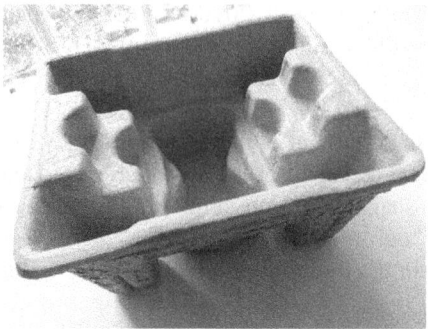

Figure 7.2 Moulded paper pulp packaging.

begin to understand the relationship between the efficiencies of packaging design and its mobility. In particular, the design of the packaging is determined by the *fit*: how tightly the camera sits within the packaging, allowing it to be protected in transportation as much as possible. The hourglass space offers strengthening to the package itself as well as space for the storage of the product peripherals such as the power source and leads.

Shifting perspective reveals more. As Figure 7.2 also demonstrates, inverting the package insert highlights the importance of the title of this chapter—why ship air? There is a tension between the overall size of the packaging, the amount of wasted air and the level of protection required. Too much packaging and the weight of the box will be increased. An increase in the size of the box will reduce the number of boxes that may be stacked on a pallet for example. But equally too little moulded paper pulp packaging and the product might be damaged. Taking a Heideggerian viewpoint that starts from the void offers a valuable means of considering this tension further. The four corner 'towers' provide an almost architectural solidity to the package. Their structure is decisive: they provide both added strength and space for housing sundry items: product leads, memory card, etc. Design is also instrumental in reducing the use of packaging materials. Again, shipping air is expensive, but so is an excess of materials (Stefansson 2008, 5). Thus the design and engineering of the package itself offers a combination of packaging protection and economy of material means. There is a symbiotic relationship between the formal spatial qualities of the product, the need for protection, the bespoke design of the packaging inserts, and the containing box. Above all, attention to packaging inserts underscores the importance of *shaping* the void that surrounds products as part of 'primary packaging'.

Filling the void: *ad hoc* packaging, dunnage and void fillers

Thus far we have seen that where the principles of 'primary packaging' and the 'product-package system' are applied as part of the product development process, there is a degree of control over the protective mobilities of the goods. However, this is far from the norm, particularly when it comes to tertiary stages of packaging and distribution (García-Arca and Prado-Prado 2008). Given the complexities and scale of global commodity mobilities it is difficult for package engineers to control the mobility of goods once they have been shipped. There needs to be a further form of packaging protection, albeit at a different scale and approach. Specifically, there is a greater need for *improvisation*. Whilst control of the void space in bespoke packaging design and package engineering is central to the protection of the goods themselves, in larger-scale distribution packaging protection is determined by a multitude of different of shapes and sizes. An *ad hoc* system becomes necessary (see Boguslaw 1965). I now consider such improvisation in relation to both the tertiary packaging scale and the protection of goods in parcel delivery networks, and explore the filling of void spaces as a form of material culture in its own right. I make a key distinction between 'shaping the void' as a piece of bespoke design determined by the shape and form of the product, and 'filling the void' as an *ad hoc* form of design (Jencks and Silver 2013). The latter in particular is reactive as opposed to proactive.

In histories of maritime freight shipping and cargo mobilities, one of the decisive problems for stevedores or longshoremen has been the challenge of loading a ship so that goods were not damaged during transit (Martin 2014). This is one of the many reasons why the shipping container became such a dominant force during the twentieth century. If we compare stacking and loading wooden barrels in a ships' hold with the cuboidal form of the shipping container, the key point is that gaps between the barrels must be filled so that they cannot move around. Shipping containers can be stacked easily on top of one another in specially designed cells on board container ships, or on dockside container stacks, but when transporting barrels and other irregular-shaped goods, the void space must be filled to protect goods from moving during shipment. The Heideggerian void again becomes pivotal: it is the space of fallibility. This is where the role of *dunnage* comes to the fore. The term has been applied to a range of situations within maritime culture, including the use of mats to protect cargo from wetting or chafing. Today, the expression "refers to any material used to protect goods or packaging from damage associated with transportation. Dunnage fills the voids between cargo, which inhibits relative movement" (Venter and Venter 2012, 467).

Historically, dunnage included sections of cordwood used to shore-up the storage of loose items of cargo in ships' holds. Today the term can be applied to a range of broader transportation contexts where goods require protection during shipment. Although containerisation resolved difficulties associated with the transportation of loose, or break-bulk, cargoes—particularly that of unnecessary damage—there is a continued problem with freight storage *inside* containers themselves. In a container full of palletised goods, the container offers outer protection, but pallets inside the container can move during transportation. The modern dunnage bag becomes key. Made from layers of special hardwearing paper and including an air valve, dunnage bags are used to fill voids between packaged goods. Where at the first level of primary packaging there is a degree of bespoke shaping of the void, at this tertiary level a more flexible, ad hoc approach is used to fill void spaces between cargo and to hold goods securely in place during shipment.[1] Although seemingly a relatively simple device, the design and development engineering of dunnage bags is critical for the safe distribution of freight around the globe (Venter and Venter 2012). For example, larger bags must endure forces of up to 42 tons during transportation. Once more we can see how void spaces in freight distribution are one of the decisive yet mundane pieces of material culture in global capitalist flows of commodities: namely a glorified paper bag.

The difference between the shaping of void spaces in package engineering and the need for greater flexibility of approach in tertiary level packaging becomes starker still when one considers the related area of 'void-fill' products used in the shipment of goods through smaller-scale delivery networks as opposed to inter-modal freight distribution. As online sales and consumer practices grow (Campbell and Ram 2016) delivery networks become ever more significant. Whilst rhetoric around new technologies such as parcel delivery drones abounds (Bamburry 2015), the reality of delivery networks and their attendant technologies is rather more banal. As with the tertiary packaging of containerisation there are important aspects around the role of packaging technologies in this context. In particular, whilst the void-fill products that often spill out over our homes as we open parcels might be seen as frustrating, throwaway things, they are critical actors in the networks of national and transnational commodity mobilities. They are also important pieces of design and material culture in their own right.

The paper in Figure 7.3 has an abstract quality that belies the everyday context of filling the void spaces in parcels. Part of this disarming quality stems from the way this simple piece of paper produces volume perceivably from nowhere. It literary fills space.

Where the package inserts discussed above are designed to fit specific goods such as consumer electronics, here—as with container dunnage bags—the materiality of this fanfold craft paper is such that it offers retail and distribution companies the ability to fill void spaces when packaging-up loose retail items. When a variety of products are placed together in cardboard boxes inevitably there will be wasted void space that must be filled in order to protect the goods. This is where the materiality of void-fill products plays a central role in smaller supply chain mobilities. Just as dunnage bags fill gaps between freight pallets, the deceptively simply mundane technology of crumpled paper does the same. Although the design thinking and package engineering behind the development of bespoke packaging solutions may appear rather more complex, the void-fill paper products themselves are still very much part of a framework of design innovation. Such innovations in the design of the packaging in Figure 7.3 stems from the transformation of a single layer of fanfold craft paper into the PaperStar™ configuration (Ranpak 2014) that provides the necessary volume of paper to protect goods. Overall, the apparent simple design of a crumpled piece of paper belies its place within wider networks of national and transnational mobilities.

Figure 7.3 Paper void-fill product.

Conclusions

This chapter has sought to illuminate the complex relationship between design and mobility. One of my central arguments is that the design of mobility (by way of logistics and supply-chain networks) takes place—in part—through the design of protective packaging itself. I have stressed the need for greater recognition of the under-represented study of protective packaging materials within the field of design cultures and mobilities studies more broadly. I also have argued for an expanded notion of packaging studies to include both aesthetically determined traditions within consumer research, and traditional discourses surrounding package engineering in academic and business-led logistics communities. These are not mutually exclusive, as my discussion of the aesthetic curiosities of packaging inserts themselves goes some way to prove. This leads to a further point of conclusion: the material culture and materiality of protective packaging design should be acknowledged for the contribution it can potentially make to the ongoing study of the material culture of supposedly mundane design artefacts.

The chapter has also strived to demonstrate the direct relationship between the development of products through the design process, and the role of protective packaging design, notably through the field of package engineering. Where the opening section argued for greater consideration to be given to the importance of packaging within the literature on commodity circuits, the following section stressed the incorporation of packaging design into product development. Packaging design is often an integrated aspect of the design process itself, whether through the direct relationship between product designers and packaging experts in the case of IKEA, or via outsourcing to a package engineering service provider. Design and mobility are parallel. As we saw with the opening example of the IKEA stool, logistical knowledge can be a central facet of their design process, albeit that recognition of the intrinsic economic acumen of a logistics-led approach to design is far from universal (Bramkley 2009; Klevås 2005).

One of the final aims of this chapter has been to direct the discussions of protective packaging away from the logistics and supply-chain literature, and towards debates within mobilities studies and design cultures. Although the work of Bello (2008) and Walker (1989) has gone some way to considering the importance of distribution and circulation to the study of design, little has been written on the place of packaging as a material form. In illuminating the *shaping* and *filling* of void spaces I have used this precise aspect of packaging to address the importance of materiality. The final sections of the chapter distinguish between the *bespoke* design of packaging inserts associated with package engineering, and *ad hoc* solutions to filling void spaces, notably through innovations in void-fill products. I would argue that there is value in considering such material forms out of context: doing so provides a productive lens through which to appreciate their curious materiality. Critically, the space *around* the product is as worthy of study as the product itself: packaging inserts and void fillers are significant registers of the relationship between design and mobility.

Note

1 For an outline of the types of dunnage bag products available see http://www.bates-cargopak.co.uk, for example.

References

Ashley, S. 1992. Handle with care: designing damage-proof packaging for products *Mechanical Engineering*, 114: 66–70.

Bamburry, D. 2015. Drones: designed for product delivery. *Design Management Review*, 26: 40–48.

Bello, P. 2008. *Goodscapes: global design processes.* Helsinki: University of Art and Design Helsinki Publications.

Birtchnell, T., Savitzky, S. and Urry, J. 2015. *Cargomobilities: moving materials in a global age.* London: Routledge.

Boguslaw, R. 1965. *The new utopians: a study of system design and social change.* Englewood Cliffs, NJ: Prentice-Hall.

Bramklev, C. 2009. On a proposal for a generic package development process *Packaging Technology and Science*, 22: 171–186.

Campbell, P. and Ram, A. 2016. Retailers grapple with online rise and ways to fill quieter aisles. *Financial Times.* http://www.ft.com/cms/s/0/eeaca352-b5fe-11e5-b147-e5e5 bba42e51.html#axzz44wWhdPwa [accessed 12 February 2016].

Crang, P. 1996. Displacement, consumption and identity. *Environment and Planning A,* 28: 47–67.

Dezeen 2015. IKEA works in a very different way to everyone else. http://www.dezeen. com/2015/02/09/ikea-design-manager-marcus-engman-interview-product-development-process-cost/ [accessed 11 February 2015].

du Gay, P., Hall, S., Janes, L., Mackay, H., and Negus, K. 1997. *Doing cultural studies: the story of the Sony Walkman.* London: Sage.

Emblem, A. and Emblem, H. 2000. *Packaging prototypes 2: Closures.* Hove: RotoVision.

García-Arca, J. and Prado-Prado, J.C. 2008. Packaging design model from a supply chain approach. *Supply Chain Management* 13: 375–380.

Hanlon, J.F., Kelsey, R.J. and Forcino, H.E. 1998. *Handbook of package engineering* (third edition). Boca Raton, FL: CRC Press.

Grier, K.C. 1988. *Culture and comfort: people, parlours, and upholstery, 1850–1930.* Rochester, NY: Strong Museum.

Heidegger, M. 1971. The thing. In *Poetry, Language, Thought.* New York: Harper Collins, pp. 163–180.

Hudson, R. 2005. *Economic geographies: circuits, flows and spaces.* London: Sage.

Hughes, A. and Reimer, S. 2004. Introduction. In *Geographies of commodity chains.* Ed. Hughes, A. and Reimer, S. London: Routledge, pp. 1–16.

Jackson, P. 1999. Commodity culture: the traffic in things. *Transactions of the Institute of British Geographers,* 24: 95–108.

Jencks, C. and Silver, N. 2013. *Adhocism: the case for improvisation* [expanded and updated edition]. Cambridge, MA: MIT Press.

Kimbell, L. 2011. Rethinking design thinking: Part 1 *Design and Culture,* 3: 285–306.

Klevås, J. 2005. Organization of packaging resources at a product-development company. *International Journal of Physical Distribution & Logistics Management.* 35: 116–131.

Knappett, C., Malafouris, L. and Tomkins, P. 2010. Ceramics (as containers). In: *The Oxford handbook of material culture studies.* Ed. Beaudry, M.C. and Hicks, D. Oxford: Oxford University Press, pp. 582–606.

Kristoffersson, S. 2014. *Design by IKEA: a cultural history.* London: Bloomsbury Academic.

Lancioni, R.A. and Chandran, R. 1990. The role of packaging in international logistics. *International Journal of Physical Distribution & Logistics Management,* 20: 41–43.

Leslie, D. and Reimer, S. 1999. Spatializing commodity chains. *Progress in Human Geography,* 23: 401–420.

Martin, C. 2014. The packaging of efficiency in the development of the intermodal shipping container'. *Mobilities,* 9: 432–445.

Martin, C. 2016. *Shipping container.* New York: Bloomsbury Academic.

Pålsson, H., Finnsgård, C. and Wänström, C. 2013. Selection of packaging systems in supply chains from a sustainability perspective: the case of Volvo. *Packaging Technology and Science.* 26: 289–310.

Porter, G. 1999. Cultural forces and commercial constraints: designing packaging in the twentieth-century United States. *Journal of Design History,* 12: 25–43.

Ranpak 2014 *FillPak M* [Product Information Sheet]. Concord Township, OH: Ranpak Corporation.

Saghir, M. 2004. *A platform for packaging logistics development – a systems approach.* Doctoral Dissertation, Department of Design Sciences, Lund University, Sweden.

Stefansson, G. 2008. *IKEA – increased transport efficiency by product and packaging redesign.* Berlin: BestLog.

Thompson, R. 2012. *The manufacturing guides: graphics and packaging production.* London: Thames & Hudson.

Twede, D. 2005. The cask age: the technology and history of wooden barrels. *Packaging Technology and Science*, 18: 253–264.

Urry, J. 2004. 'The 'system' of automobility'. *Theory, Culture & Society*, 21: 25–39.

van den Berg-Weitzel, L., and van de Laar, G. 2001. Relation between culture and communication in packaging design. *Brand Management*, 8: 171–184.

Venter, M.P. and Venter, G. 2012. Overview of the development of a numerical model for an inflatable dunnage bag. *Packaging Technology and Science*, 25: 467–483.

Vidler, A. 2001. *Warped space: art, architecture, and anxiety in modern culture.* Cambridge, MA: MIT Press.

Virilio, P. 2006. *Negative horizon.* London: Continuum.

Walker, J.A. 1989. Production-consumption model. In *Design history and the history of design*. Ed. Walker J.A. and Attfield, J. London: Pluto Press, pp. 68–73.

8 Designing signals, mediating mobility

Traffic management and mobility practices in interwar Stockholm

Martin Emanuel

Introduction

On 23 January 1925, Stockholm's first traffic lights became operational.[1] The four lampposts, one in each corner of the busy downtown intersection Kungsgatan-Vasagatan, had two colours only (red/green) and were manually operated by traffic police officers trained by the Swedish manufacturer AGA (see Figure 8.1). The novelties were praised as a success in the daily papers ("Succés för den optiska trafikpolisen" 1925; "De optiska signalerna" 1925; "Ljussignalerna på Kungsgatan fungera bra" 1925). By 1935, twelve intersections in Stockholm were operated using traffic lights. During this ten year-period, experiments were made with different phasing and colour combinations. The first signal installation with both manual and automatic direction was put into operation in 1930. By 1933 a particular "Stockholm" phasing system had been established using red, green and yellow lights. Among twelve installations active in 1935, one could only be operated manually (by traffic constables); one only automatically; and ten could be operated either way (Dufwa 1985, 88–90; see also Carlén 1936).

During the following decades, traffic lights in Stockholm were further developed. In 1947 the first vehicle-induced traffic lights were tested, making it possible for vehicles to physically activate the green light during low traffic. From 1953 traffic lights were coordinated to allow "progressive" signal regulation. In this way, commuter traffic on major streets had a "green wave": vehicles could pass through several signal-regulated crossings without stopping for a red light. Traffic lights began to be coordinated in more sophisticated ways, using feedback mechanisms, and from 1966 computers were used to optimise coordination in a complete area to reduce total waiting times (Dufwa 1985, 90–94).

Traffic lights are one of many innovations in a century-long attempt to control the flow of, and come to grips with the risks involved in, urban traffic. They are also part of a system "that attempts to impose a strong social control over the most fundamental of human behaviour, whether to move or to be still" (McShane 1999, 379). At the same time they are an integral and obvious part of our everyday technological culture and something upon which road users hardly reflect. Like many other urban technologies, they are "second nature", and also indispensable; we hardly notice them until they fail (Edwards 2003).

Figure 8.1 Stockholm's first traffic signal.

Existing work on traffic control is characterised by, on the one hand, valuable historical studies into the details of standardisation of traffic light design and the views and actions of different professional groups involved in the management of interwar urban mobility (McShane 1999; Buiter and Staal 2006; Norton 2008). On the other hand we have the thought-provoking essays of Bruno Latour (1992, 1994) with hypothetical observations regarding the agency of speed bumps or the scripting of seat belts; that is, how material objects interfere with human action and interaction. The central focus of this chapter is upon the (partial) shift from police-controlled to machine-controlled mobility in interwar Stockholm. I begin by reflecting upon what has been termed the "traffic crisis" of the interwar period, then consider the process of technology and knowledge transfer in the introduction of traffic control measures in Stockholm, and evaluate the piecemeal transition from hand-signalled to mechanised control of urban mobility. Drawing upon archive material, newspaper articles and published material related to the introduction and early development of traffic lights I aim to understand what lights "do" in a Latourian sense. The chapter thus seeks to highlight design as a political and interest-driven domain, and the role of artefacts in producing mobility.

Traffic control comes to town

Nineteenth century traffic was essentially unregulated. Road users typically managed by customary conduct, such as keeping to the right, and traffic conflicts were solved informally. Slow speeds and short stopping distances meant that vehicle priority did not have to be defined and was usually "by might rather than by decree" (Lay 1992, 28). Increasing traffic at greater speeds in the first two decades of the twentieth century led to more serious traffic accidents, congestion and chaotic streets. While this "motor traffic crisis" took place in the 1910s in American cities, many European cities experienced a similar crisis in the mid-1920s—although it should be acknowledged that such challenges, or rather the perceptions of them, appear to be fundamental to urban life and thus not new to these decades (Norton 2008, 47–49; Lay 1992, 84; McShane 1999, 380–381; Buiter and Staal 2006; Weinstein 2006).

In response to such "crises", experts of different sorts innovated in traffic control with the aim to control road users and mobility practices. The development of traffic lights and other innovations in traffic control is not solely a national story but one that requires transnational framing. McShane (1999) argues that the American experience was crucial in the development of traffic lights around the globe. When non-US cities began to draw on American experiences, they accepted a system of uniformity (McShane 1999, 389–395). Studying the introduction and diffusion of traffic lights in the Netherlands, however, Buiter and Staal (2006) stress the importance of local experimentation of traffic police departments and Dutch and European traffic light suppliers in the great variety of designs in Dutch cities. They credit the Dutch government and national and international organisations for the standardisation of three-coloured traffic lights during the 1930s, based on European rather than American experiences (Buiter and Staal 2006).

The Swedish example tells an important story of transatlantic circulation of knowledge and technology rather than of linear transfer or a European-only endeavour. In response to a surge in traffic, and motorised traffic in particular, in the 1920s, measures were introduced to facilitate traffic and improve traffic safety in Stockholm: special traffic islands for tramway halts; one-way traffic regulations and roundabouts; road lanes, pedestrian crossings and parking spaces for cars; and traffic signs, silent policemen and traffic lights (Bergman 1929, 10; Dufwa 1985, 82 and 108.)

The Swedish company AGA (Aktiebolaget Gasaccumulator) played a crucial role in the introduction of silent policemen and traffic lights in Sweden. Founded in 1904, the company revolutionised lighthouse technology and had its international breakthrough when it won a contract to set up lighthouses along the Panama Canal in 1912. Increases in road traffic in the US presented AGA with a new, international market for its acetylene-based lighting. In 1919, the American subsidiary American Gas Accumulator Co., founded in 1911 to facilitate AGA's US activities, set up a special Traffic Engineering Division in 1919 and hired as its manager Guy Kelcey, mechanical engineer and later a founding member and charter Director of the American Institute of Traffic Engineers (ITE 2015).

In 1930, AGA delivered 4,500 silent policemen to US cities and another 3,000 for use on American roads.[2] The company delivered their first traffic lights in the US in 1924—only a year before Stockholm's first traffic lights—and within a ten-year period they had helped to spread the novelty across the country.

Through the interwar period, AGA delivered most silent policemen and traffic lights used in Swedish cities (Bergman 1929, 25). AGA's US experience was important both in terms of knowledge and technology transfer and as a selling point for Swedish cities and traffic police departments. The best evidence of the effectiveness of silent policemen was, according to the company, their spread in the US. An *Aftonbladet* motor journalist agreed: "This marvellous Swedish invention has had a serious breakthrough in America and should for this reason be considered valuable also in our country" ("Var behövs …" 1924). Although not necessarily a conscious strategy, AGA's involvement in the introduction of traffic control devices in Sweden—where the Swedish parent company could build on competencies brought back from their American subsidiary—illustrates the key role of transnational firms in the circulation of knowledge and technology in the interwar period.[3]

De-humanising traffic control

Across both the US and Europe, the introduction of traffic lights can be regarded as a first step in replacing the physical, direct supervision by traffic police with more automatic and computerised control measures based on traffic engineering. In the US, Norton (2008) argues that traffic engineers gradually replaced policemen as the most important professional group from about 1920. The initial establishment of traffic codes, street markings, one-way streets, traffic circulations, semaphores and traffic towers, was complemented with traffic lights and restrictions on kerb parking (Norton 2008, 54–64, 134–146).

By the mid-1920s, major American cities adopted traffic lights with automatic timers. While replacing expensive traffic police, automatic controls required professional electricians to maintain them and careful engineering surveys of traffic to set appropriate cycles, since both were beyond the skills of most police departments. The progressive traffic signal system, first installed by General Electric in Washington, DC in 1926, needed careful adjustment by engineers, involving the estimation of average speeds, acceleration and retardation times, and often a long trial-and-error procedure. Automatic controls increased commuting speeds considerably and the system spread quickly in the US (McShane 1999, 385–388).

In this process traffic police lost their previous role as traffic regulators; and their responsibilities became more educational—for example, to socialise road users into proper traffic behaviour through educational campaigns (McShane 1999, 391). Nonetheless the transition from police-control to engineered control was not immediate. In the US, the traffic police, manufacturers of traffic signals, and downtown businesses—not municipal engineers—devised the first traffic light systems. And when European cities adopted traffic lights the first installations were all based on manual control by policemen (McShane 1999, 382–385; Lay 1992, 187–188).

The initiative to install the first traffic lights in Stockholm in 1925 came from the police.[4] The lights were designed in close collaboration with AGA: the six police officers managing the lights were trained by the company ("Ljussignalerna på Kungsgatan fungera bra" 1925). Prompted by positive experiences from the lights in Kungsgatan-Vasagatan, the traffic police urged city authorities to install similar ones in two other intersections in the city centre. The city's traffic committee approved one installation but wanted to delay the second because of cost considerations.[5] No further lights were installed during in the 1920s: Stockholm instead developed simpler, cheaper tools to ease the work of the traffic police (Sjöberg 1928, 219–224).

Still, the traffic police kept pushing for the installation of more traffic lights, and by the early 1930s successfully so. Their arguments were manifold: to cut costs; improve the work situation for police officers (who found it difficult to oversee the whole traffic situation, and ran great risks by standing in the midst of intense traffic); and to release them for other tasks such as traffic supervision.[6] A traffic police officer doing service during the transition to mechanised traffic direction remembered it as more comfortable to manage the mechanised device than to "stand for hours and 'wave hands'" (Cassel 2001, 153). Stawström argued in 1935 that the choice between an officer's hand signals and optical signals was ultimately "a matter of costs", which usually favoured optical signals.[7] Two years earlier, at a conference gathering the police administrations in Stockholm, Gothenburg and Malmö, participants agreed to "cherish the technical system because of its economic merits". In less than one year, they argued, one traffic light installation saved the annual cost of one traffic officer ("Trafiksignal amorteras på ej fullt året" 1933).

In the US during the first half of the twentieth century, traffic direction was mechanised, automatised and coordinated. A new professional group with new means and ends of traffic management took command. New practices of traffic control were driven by actors who responded differently to increasing automobility. By contrast, in Stockholm, the interwar period saw only initial mechanisation and only partial automatisation: police officers still operated most lights, although lights could be automatically operated as well. In fact, the traffic police, together with signal manufacturers, stand out as the driving force in the process. Before reflecting in more detail upon the effects of this partial transition on mobility practices and considering the specific impact of traffic lights as socio-technical systems, I briefly outline an actor-network-inspired approach to understanding traffic control.

Actor-network theory (ANT) and traffic lights

Technology and the social context cannot be separated but are mutually constitutive. Our material world is a product of social processes at the same time as technology brings order, shapes and constitutes our social existence in terms of practices, identities, norms, discourses and institutions (Jasanoff 2004, 2–3). Though traffic is commonly understood as highly decentralised, impersonal and based on autonomous road users, traffic management constitutes a framework for behaviour and interaction; traffic control may be considered a way of "engineering" interaction (Normark 2006, 12–15). As Ogborn (2000, 264) argues for late twentieth century signal-regulated intersections, "[d]rivers and lights are meshed together as the moving fabric of the networked city".

How did interaction between road users change through the use of hand signals of the traffic police and later by automated traffic lights for guidance of traffic? Reconstructing hypothetical situations, the presence of a police officer should have meant that road users, rather than negotiating and coordinating conduct at each street crossing, relied more upon the officer and less upon interchange with other road users. Their centre of attention ought to have turned towards the police officer. In the case of replacing the officer with automatic traffic lights, two-way communication between road users and police officers should have been replaced by one-way communication from signal to road user only. However, for a more detailed understanding of the effects of traffic lights on human interaction and mobility practices we may approach the question equipped with the toolbox provided by actor-network theory. Rather than suggesting that "stuff" has its own intentions—a common misunderstanding about ANT—it proposes that that materials "do" things through their associations within a network of actors, human and non-human. Since it has an effect on the outcome of an event it is appropriate to speak of the agency of things (Latour 2005).

Inspired by actor-network-based approaches to the study of architectural design (Yaneva 2009) ANT appears useful for understanding the "distributed agency" of the road user–technologies–designer, or the "relational materiality" (Law 1999) at stake in urban traffic. Was, for example, the introduction of traffic lights a mere replacement and mechanisation of the assignment of traffic police officers? Or did changes (intentional/unintentional/unpredictable/unwanted) occur in the transition? The language used in interwar Stockholm sometimes suggests straightforward replacement. Traffic lights were initially termed "optical traffic police" in Swedish newspapers, and in 1936 a Swedish traffic signal expert of referred to them as a "mechanisation of police supervision" (Carlén 1936, 21).

I suggest differently. The agency of artefacts is most easily discernible in the early innovation phase of a technology, as associations are redistributed through a heterogeneous network. In this stage, artefacts are "mediators", while they shortly thereafter fade into the background as "intermediaries" (Latour 2005). This does not mean that they no longer "act", rather that agency is no longer visible. Innovation processes are thus favoured items of analysis in ANT scholarship. If we are interested in technologies introduced in the past, the historical method proves necessary—albeit that it is challenging to accomplish the "thick descriptions" of ethnomethodology that are characteristic for ANT. In the following section I recapture the local innovation process of traffic lights and the different assessments of hand and machine-based traffic signals in Stockholm. I also refer to British handbooks and manuals available in Swedish libraries in order to reveal the relation (as imagined by the designers) between artefacts and users.[8]

According to Akrich (1992) designers make assumptions about the potential users of a technology—their preferences, competencies, motives and aspirations—and then "inscribe" these in the technical content of the object. Such "scripts" appear as the end-product of an operationalisation of the designer's perception of the (appropriate) relationship between the technology and the user, laid down in urban infrastructure. Through these "scripts", experts "prescribe" appropriate forms of mobility (Akrich 1992, 207; Akrich and Latour 1992; see also Woolgar 1991). To "de-scribe" something is then the backward process in relation to the designer's inscription of the script in the artefact; it is what the analyst does to read the script from an artefact and the network it is associated with. My ambition below is to de-scribe traffic lights.

De-scribing traffic lights

In de-scribing traffic lights, articles in Swedish press on the limits of police officers' hand signals are as revealing as those hailing optical signals as a measure of modernity. In a series of articles in the mid-1920s, *Aftonbladet*'s well-known motor journalist Carl Skånberg raged against police officers' hand signals (Transmitter 1923; Skånberg 1925). In particular he argued for the reintroduction of mandatory police signs (and absolute obedience of road users), which had been dismissed in Stockholm's new 1923 ordinance. According to Skånberg (Transmitter 1924a), this would not only yield more accidents; but also it would bring general "demoralisation", since it gave road users "an adequate excuse for not obeying a traffic signal". When police signs were mandatory, the driver could direct his attention away from traffic to the constable and "wait for his signal and drive on the responsibility of the police" (Skånberg 1927)—thus confirming the hypothetical argument above, about the impact of police-control on road user interaction. If the combination of mandatory police officer signs and modern traffic meant that constables had to "wave arms like windmills", Skånberg argued, then perhaps it was about time to think about mechanical assistance to direct traffic (Transmitter 1924a).

Criticism of the traffic police was manifold. Often it was class-based and constructed traffic constables as young, arbitrary, impolite and authoritarian ("corporal-like") beyond the justifiable (Stillman 1928; Kleen 1932). Constables' hand signals also were disparaged for lacking clarity and consistency. In 1930, one critic found the hand signals "unclear or difficult to understand", and he speculated that it had to do with the "shyness among some constables", who did not want to stand out in a crowd more than necessary ("Bättre signaler ..." 1930). The same year, Axel Norlander, general secretary of the Swedish Royal Automobile Club, similarly noticed how officers often only nodded or waved disinterestedly at stomach-height to road users to proceed. Not only should hand signals be clear, they also needed to be standardised: "regulated, determined and distinct, and it must be obeyed without hesitance" ("Trafiksignalerna otillfredsställande ..." 1927).

The local press also commented upon differences between policemen's hand signals and mechanised traffic lights. Stockholm's first lights were seen to have a precision that allowed intersections to "swallow" more traffic ("Succés för den optiska trafikpolisen" 1925). Their instruction was "rigid, but clear", in contrast to the policemen's instruction. Much traffic made it difficult for road users to see a traffic police officer from a distance. Lights at the Vasagatan-Kungsgatan intersection, however, could be seen from hundreds of metres away, which gave road users plenty of time to judge how to behave once they reached the intersection. In the same paper, one driver reported that it was considerably easier to keep an eye on the traffic lights than to pay attention to a "windmill-turning" officer on the street corner ("De optiska signalerna" 1925). Efficiency, clarity and visibility were seen to be important qualities of manually operated traffic lights in contrast to the hand signals of traffic police.

As to their impact on mobility practices, the first assessments in newspapers differed. "It was striking", *Svenska Dagbladet* noted,

> how during the complete day, the cars that had signal to go, rushed through the intersection. In other words, speeds were higher than usual. This is of course only a merit, since traffic is so heavy that it does not allow unnecessary delays, but it is on the other hand extremely important for the pedestrians to observe. For while the new system has the ability to safely guide the pedestrians over the roadways, it also implies greater danger for those who defy or neglect its warnings.
>
> ("Succés för den optiska trafikpolisen" 1925)

In contrast to the claim about higher speeds, *Stockholms-Tidningen* reported slower speeds than usual, given the inexperience among road users to the new circumstances ("Ljussignalerna på Kungsgatan fungera bra" 1925). Similarly, *Aftonbladet* recorded a "certain hesitance among several road users", in particular the horsemen ("God debut för trafiksignalerna" 1925).

Stockholm's early traffic lights could commonly be operated both manually and automatically. A 1936 Swedish motor magazine booklet explained differences between the two types of signals:

> Traffic lights are cheaper than police officers for directing traffic at intersections and they can alternately stop and start the traffic stream more quickly than an officer is able to. They are suited for different amounts of traffic and can, under appropriate circumstances, with advantage replace traffic officers except at such instances, when personal judgment is required.
>
> (Carlén 1936, 14–15)

Here, efficiency and economy spoke in favour of automatic traffic signals, while lack of personal judgement could be a drawback. In fact, as late as 1935, the traffic police and Stockholm city authorities agreed that in some instances, manual traffic direction could be more flexible than optical signals, and even admit higher traffic capacity.[9] While there was a renewed interest in traffic signal systems after the Second World War, forms of signal regulation were strongly criticised. The drawbacks of manually operated signals were considered to be their high operating cost as well as difficulties for policemen to correctly judge phases in accordance with the highest possible traffic capacity and comfort for road users. The fixed phasing of automatic lights, on the other hand, could satisfy the road users at peak hour traffic, but in return made them wait for a green light even when there was no crossing traffic (Dufwa 1985, 90). This would eventually lead to the preference of traffic-induced lights.

In the US, automatic traffic lights also were criticised for lacking personal judgement. There were also concerns as to whether people would obey a "dumb" machine. This was an important reason why manual operation was common long after automatic solutions became available (McShane 1999, 382–385). The first lights were all based on manual control by policemen, since it was believed that road users would ignore them unless they saw a police officer (Lay 1992, 187–88).

Handbooks and manuals for traffic lights reveal more detailed expert scripts in terms of appropriate road user behaviour. Since no such handbooks in Swedish are available I make use of British handbooks which were available in libraries in Stockholm. Despite differing national contexts, the British handbooks parallel evidence from Swedish sources. For example, Harrison and Priest (1934, 2–3) include a lengthy list of reasons why automatic traffic lights were superior to the hand signals of the police. A number of the "failings" of traffic officer control parallel critiques levelled in Swedish newspapers. Traffic officers were considered expensive, could become tired and could not cope with the same levels of complexity as traffic lights. Hand signals were not as visible, clear or exact as traffic lights. However, Harrison and Priest (1934) also emphasise the effects of hand signals upon road users. They suggest that when confronted by a police officer, road users became hesitant or timid. Human flexibility also could create ambiguity: for example, a police officer could be "prone to wait for stragglers instead of dealing promptly with waiting traffic".

It should be noted, however, that the perspective of traffic engineers like Harrison and Priest (1934) was not entirely unchallenged. Skilled in traffic engineering, assistant commissioner of police at Scotland Yard, Alker Tripp (1950 [1938], 258–259), agreed with most of the positive effects of the transition from police-control to automatic control. Automatic lights had advantages of higher endurance, better capability to work on shorter cycles, precision, better possibility for coordination, economy, and greater visibility. Tripp (1950 [1938]) even agreed with the engineering perspective that officers implied a lack of uniformity, as well as vagueness and hesitation on the side of the road user. In his view, however, these advantages needed to be carefully weighed against their drawbacks.

In a seven-point list of disadvantages (Tripp 1950 [1938], 259–260) many refer to the relative inflexibility of automatic lights: they could, for example, not effectively handle individual right-hand turns, pilot individual pedestrians across the road, or give right of way to emergency vehicles. Most importantly, however, lights could not replace, but only complement officers: particularly in the case of signal failure, "whereas Police Officers are perfectly well able to regulate traffic without the aid of signals, signals are not correspondingly able to regulate traffic without the aid of Police Officers" (Tripp 1950 [1938], 258). Most interesting are references to the importance of enforcement and the over-confidence of road users:

1. Signals can only give direction and cannot enforce them. If their directions are disregarded, the dangers at a junction are likely to be increased instead of lessened, because traffic that has been given the green signal is proceeding with confidence – and generally at a higher speed than that at which it would otherwise have approached the junction.

4. The 'confidence and quickness' of drivers in response to automatic signals is liable to develop into over-confidence and disregard (a) by their failing to pull up when directed, though well able to do so, (b) by their stopping suddenly if they have approached the green signals at speed and it has changed to amber; and (c) by their starting before the full release signal is given.

Though valuing them differently, Tripp's (1950 [1938]) assessment strengthens the thesis that traffic lights made road users too hesitant. The exchange between British traffic engineers and traffic police officers reveals the prescriptions of traffic lights; that is, what an artefact allows or even forces a given actor to do or prevents him/her from doing. Traffic engineers maintained that the traffic lights made road users less hesitant and timid. Traffic police agreed but valued the difference differently: for them, a lack of hesitation and over-confidence risked creating dangerous situations when people blindly followed traffic lights.

Both engineers and traffic police suggested that traffic lights had a complexity-reducing function, freeing drivers from hesitation. Traffic lights inoculated—in ANT vocabulary, they prescribed—an on–off behaviour among road users, in contrast to continuous negotiation between them, or between them and the officer. The on–off behaviour was further fed by associations made in advertisements and elsewhere, between a traffic light about to change and the starting signal of a motor race. In 1926, for example, the company AB Amerikanska Motor Importen (American Motor Import) equated urban driving with motor sport: "Why be left behind when the traffic light blinks, when the Champion spark plug allows an immediate take-off and takes you to the lead at once!" (*Svenska Dagbladet*, 11 June 1926, p. 13).

The transition towards clarity and exactness, in opposition to vagueness and hesitation, brought about by traffic lights, is assumed to reduce the number of possible actions and thus facilitate the decision-making process. There is no room for hesitation. Traffic lights prescribe that road users do not drive against red lights—the dangers involved if they do make it almost impossible.[10] While it may indeed also be problematic to drive against the clear hand-signal of a traffic police officer, human regulation entails a greater degree of flexibility and somewhat more room for negotiation. Put differently, traffic lights delegated the responsibility to comply with traffic rules to road users: As Ogborn (2000, 263) suggests, "the police officer is also installed within us. ... We construct our urban liberties through the discipline of the lights".

Ogborn (2000, 264) also points to a limitation of the analysis made here. While traffic lights "recognize each driver as an equal, individualized and independent citizen", this state "is undercut by the property qualification of car ownership". Indeed, traffic lights were not fully neutral in relation to different means of transportation. McShane (1999, 386) puts it bluntly for the American case: "Traffic engineers treated pedestrians as second-class citizens". Since signal timings in coordinated systems were based on vehicle speeds they helped to redefine streets as motor thoroughfares where pedestrians did not belong (Norton 2008, 136–138).

Traffic lights prescribed pedestrians to be careful and to give priority to other traffic. Users do not, however, necessarily accept, or subscribe to, the (imagined) usages scripted by experts (Oudshoorn and Pinch 2003; Trentmann 2009). In the case of Stockholm, pedestrians were considered a particularly problematic road user group ("Ljussignalerna på ..." 1925). Beginning in the late 1920s, Stockholm experimented with different ways of enabling pedestrians to more smoothly and safely cross the street (Bergman 1929, 19–20), including the use of sound signals. Whether they lacked the competence or the will to adhere to traffic lights, such initiatives suggest that pedestrians adopted "antiprograms" to the scripts implemented through traffic lights, which led to "re-inscriptions" of the technology. Although more detailed scrutiny of the influence of pedestrian antiprograms upon urban mobility systems is beyond the scope of this chapter, it provides an intriguing subject for further enquiry.

Conclusion

Drawing upon existing scholarship on the history of traffic signals and complemented by a detailed study of interwar Stockholm, this chapter has revealed how technologies for traffic management were introduced in Sweden through a circular transatlantic process under the guidance of the Swedish electromechanical firm Aktiebolaget Gasaccumulator, AGA. Based on experiences with other lighting technologies, the company's US subsidiary also managed to seize parts of the American market for street traffic control technologies. This was in turn of symbolic importance (USA as the future) and provided competitive advantage in terms of knowledge for capturing and indeed cocreating with local police and traffic departments the nascent Swedish market for traffic lights.

The chapter demonstrates that traffic lights did not merely replace what previously had been the assignment of the traffic police. Important transformations of communication and mobility practices occurred in the transition from human to mechanised guidance of traffic. While there was indeed great ambivalence towards "the human factor" in police-direction, and different assessments about the pros and cons of traffic lights, there was consensus among contemporary actors that lights brought clarity to and less hesitation in urban traffic. Traffic lights prescribed an on–off behaviour among road users, which could be broken only at great risk. Lights also delegated the responsibility for complying with traffic rules to road users themselves.

Traffic lights were introduced in the interwar period as mediators of human interaction and mobility practices—whether one characterises the development of such technical systems as progress and sophistication or as part of a process in which people are forced to subsume to technology. A central claim of this chapter, however, is that new traffic control technologies are not a straightforward and "natural" result of increasing traffic, congestion and accidents. While such developments may be important contexts, the case of traffic lights illustrates the crucial importance of an assemblage of human and non-human actors. The authority and power of the traffic police is evident through their physical appearance and conduct. The power of traffic control systems is less apparent, but it is incorporated in the materialities of which it is made up—and as least as strong.

Notes

1 In this chapter, "traffic lights" are used instead of "traffic signals" to emphasise the difference between optical traffic signals and the hand signals of traffic police.
2 This and all following details of AGA are drawn from the Archive of Aktiebolaget Gasaccumulator, held at the Centre for Business History, Stockholm and particularly F 16 D, Volumes 1–5, "Trafiksignaler Trycksaker 1920–1940-tal".
3 Scholars studying late twentieth century multinational corporations argue that these arise due to their superiority in organising knowledge transfer across national borders (Kogut and Zander 1993).
4 Gatukontorets arkiv, held at the Stockholm City Archive (hereafter GA), EI: D. nr. 1924/24. "P.M. med anledning av Polismästarens hemställan om utförandet av signalanordning i gatukorset Kungsgatan–Vasagatan samt på grund därav inkomna anbud", 17.11.1924; Gatukontoret till Gatunämnden 1.12.1924; Överståthållaren till Stockholm stads gatukontor 6.10.1924.
5 Stockholm City Archive, Överståthållarämbetet, Polisärenden 5 (hereafter ÖÄ5), Trafikinspektörens expedition, CI: D.nr. 797/25. ÖÄ till Gatunämnden 16.6.1925; GK-tjut 18.6.1925; GN-protokoll 1.7.1925.
6 GA, EI: D.nr. 2039/30. Gatukontoret till Gatunämnden 13.4.1931, "Ang. ändring av signalsystemet i gatukorset Kungsgatan-Vasagatan", D.nr. 900/31. ÖÄ (Eric Hallgren) till Stockholms stads gatunämnd, 8.5.1931; D.nr. 941/31. ÖÄ (Eric Hallgren) till Stockholms stads gatunämnd, 21.5.1931. ÖÄ5, Trafikinspektörens expedition, CI D. nr. 128/1932. Trafikinspektör Stawström till polismästaren, 22.2.1932.
7 GA, EI: D.nr. 2584/35. "Yttrande", Trafikinspektör Stawström, 13.1.1936.
8 These are two of several strategies proposed by Latour (2005, 79–82) to get artefacts to "speak"—to reveal what they make other actors, human and non-human, do—if the innovation process is not immediately accessible for review.
9 17 GA, EI: D.nr. 2584/35. "PM till Polismästaren", Trafikinspektör Stawström, 19.11.1935; Gatunämnden till ÖÄ 18.12.1935; "Yttrande", Trafikinspektör Stawström, 13.1.1936.
10 It is important to note potential requirements for re-inscription: in response to on–off behaviour, a (yellow) third light was introduced in the 1930s to prepare drivers for signal change and as a period to clear the intersection of traffic before allowing other streams to proceed.

References

Akrich, M. 1992. The de-scription of technical objects. In *Shaping technology/building society: studies in sociotechnical change*. Ed. W. E. Bijker and J. Law. Cambridge, MA: MIT Press, pp. 205–224.

Akrich, M. and Latour, B. 1992. A summary of a convenient vocabulary for the semiotics of human and nonhuman assemblies. In *Shaping technology/building society: studies in sociotechnical change*. Ed. W. E. Bijker and J. Law. Cambridge, MA: MIT Press, pp. 259–264.

"Bättre signaler från trafikpoliserna". *Aftonbladet* 21.7.1930.

Bergman, C. G. 1929. *Trafiken och gatorna i Stockholm*. Stockholm: Nordisk bokhandel.

Buiter, H. and Staal, P. E. 2006. City lights: regulated streets and the evolution of traffic lights in the Netherlands, 1920–1940. *Journal of Transport History* 27: 1–20.

Carlén, H. 1936. *Bilismens signalsystem: signaler och säkerhetsanordningar för gatu- och vägtrafik: handbok för polismän, gatu- och vägmyndigheter, bilskolor, motorförare, cyklister m. fl.* Stockholm: Svensk motortidning.

Cassel, G. 2001. Kriminalpolistjänst i tiden. In *Polistjänsten inifrån: minnesanteckningar från polistjänsten under 1900-talet*. Ed. L Silverbark. Stockholm: Polisveteranerna i Stockholms län.

"De optiska signalerna gjorde lyckad debut. Trafiken går ledigt". *Dagens Nyheter* 23.1.1925.

Dufwa, A. 1985. *Stockholms tekniska historia: trafik, broar, tunnelbanor, gator.* Stockholm: Kommittén för Stockholmsforskning.

Edwards, P. N. 2003. Infrastructure and modernity: force, time, and social organization in the history of sociotechnical systems. In *Modernity and technology*. Ed. Thomas J. Misa, P. Brey and A. Feenberg. Cambridge, MA: MIT Press.

"God debut för trafiksignalerna vid Vasagatan". *Aftonbladet* (Stockholmsupplagan) 22.1.1925.

Harrison, H. H. and Preist, T. P. 1934. *Automatic street traffic signalling*. London: Pitman.

ITE. 2015. *Guy Kelcey Institute of Transportation Engineers*. Available at http://www.ite. org/aboutite/honorarymembers/KelceyG.asp, accessed 25.3.2015.

Jasanoff, S. 2004. The idiom of co-production. In *States of knowledge: the co-production of science and social order*. Ed. S Jasanoff. London: Routledge, pp. 1–12.

Kleen, E. 1932. Trafikkonstapeln är icke mer än en människa. *Stockholms-Tidningen* 4.5.1932.

Kogut, B. and Udo, Z. 1993. Knowledge of the firm and the evolutionary theory of the multinational corporation. *Journal of International Business Studies* 24: 625–645.

Latour, B. 1992. Where are the missing masses? In *Shaping technology/building society: studies in sociotechnical change*. Ed. W. E. Bijker and J. Law. Cambridge, MA: MIT Press.

Latour, B. 1994. On technical mediation: philosophy, sociology, genealogy. *Common Knowledge* 3: 29–64.

Latour, B. 2005. *Reassembling the social: an introduction to actor-network-theory*. Oxford: Oxford University Press.

Law, J. 1999. After ANT: complexity, naming and topology. In *Actor network theory and after*. Ed. J. Law and J. Hassard. Oxford: Blackwell.

Lay, M. G. 1992. *Ways of the world: a history of the world's roads and of the vehicles that used them*. New Brunswick, NJ: Rutgers University Press.

"Ljussignalerna på Kungsgatan fungera bra". *Stockholm-Tidningen* 23.1.1925.

McShane, C. 1999. The origins and globalization of traffic control signals. *Journal of Urban History* 25: 379–404.

Normark, D. 2006. *Enacting mobility: studies into the nature of road-related social inter-action*. Göteborg: Section for Science and Technology Studies, Sociology Department, Göteborg University.

Norton, P. D. 2008. *Fighting traffic: the dawn of the motor age in the American city*. Cambridge, MA: MIT Press.

Ogborn, M. 2000. Traffic lights. In *City A-Z*. Ed. S. Pile and N. Thrift. London: Routledge, pp. 262–264.

Oudshoorn, N. and Pinch, T. (Eds) 2003. *How users matter: the co-construction of users and technologies*. Cambridge, MA: The MIT Press.

Sjöberg, S. 1928. *Signaler i samfärdselns tjänst*. Stockholm: Geber.

Skånberg, C. 1925. Det bättrar sig. *Aftonbladet* 2.3.1925.

Skånberg, C. 1927. Konsekvens! *Aftonbladet* 9.4.1927.

Stillman, 1928. Stopp! På nio olika språk. *Svenska Dagbladet* 8.7.1928.

"Succés för den optiska trafikpolisen". *Svenska Dagbladet* 23.1.1925.

"Trafiksignal amorteras på ej fullt året". *Svenska Dagladet* 20.5.1933.

"Trafiksignalerna otillfredsställande. Ett reformförslag". *Svenska Dagbladet* 19.1.1927.

Transmitter. 1924a. Stockholmspolisens trafiksignaler. *Aftonbladet* 21.6.1924.

Transmitter. 1924b. Trafikkultur i Sverige. *Aftonbladet* 13.9.1924.

Transmitter. 1923. Trafiksignalerna. *Aftonbladet* 6.10.1923.

Trentmann, F. 2009. Materiality in the future of history: things, practices, and politics. *Journal of British Studies* 48: 283–307.

Tripp, H. A. 1950 [1938]. *Road traffic and its control*. London: Edward Arnold.

"Var behövs trafikfyrar?". *Aftonbladet* 20.12.1924.

Weinstein, A. 2006. Congestion as a cultural construct: the 'congestion evil' in Boston in the 1890s and 1920s. *Journal of Transport History* 27: 97–115.

Woolgar, S. 1991. Configuring the user: the case of usability trials. In *A sociology of monsters: essays on power, technology and domination*. Ed. J. Law. London: Routledge, pp. 58–99.

Yaneva, A. 2009. *The making of a building: a pragmatist approach to architecture*. Bern: Peter Lang AG.

9 MotoGP and heterogeneous design

Philip Pinch and Suzanne Reimer

Introduction

This chapter takes as its central focus the design of the MotoGP racing motorcycle. Our key aim in scrutinising MotoGP is to conceptualise the location of design and its meaning: to reflect upon where and how design happens; and to enrich understandings of the heterogeneous, networked and distributed character of design. Whilst many consumer objects are mobile, our analysis extends beyond the design of the motorcycle as a single mobile artefact to consider both the motorcycle–rider assemblage (Pinch and Reimer 2012) as well as the complex architecture of design technologies which support MotoGP racing. We stress the agency of material objects and non-material inputs (expressions and emotions) that operate relationally and act upon the MotoGP motorcycle, crucially defining its movement and performance. The chapter emphasises the multifaceted nature of design: it runs through mechanical engineering; the design of electronic control systems; the regulation of international motorcycle racing; and the physical body of the rider.

The world championship MotoGP competition is characterised by its sponsors, Dorna Sports Ltd., as

> an eighteen-race series visiting over a dozen countries, multiple continents and with pan-global television coverage. Upwards of ten nationalities of the world's most skilled riders line a grid armed with cutting-edge motorcycle technology with prototype machinery fielded by manufacturers including Aprilia, Ducati, Honda, Suzuki and Yamaha.
>
> (2016a)

The race weekend begins on a Friday, and involves a series of 'free practice' and warm-up sessions scheduled over three days, culminating with the 'top class' MotoGP race on a Sunday.[1] Although overseen by the global organisation Fédération Internationale de Motocyclisme (FIM), MotoGP is directly regulated by the international sports management, marketing and media company Dorna, headquartered in Madrid (2016b). At the time of writing Dorna specifies all racing, qualifying and practice rules as well as technical regulations; it is also the exclusive MotoGP commercial and television rights holder.

Whilst it may appear to represent an extraordinary performance of motor-cycling, in practice the apparatus of MotoGP is integrally related to production motorcycles purchased by individual consumers. In contrast with Formula 1 car racing, in which vehicles differ dramatically—both technically and visually—from road cars, MotoGP motorcycles resemble everyday sports motorcycles and use similar technologies (Cameron 2009, 9). Technical evolution in MotoGP, such as the development of electronic traction control systems, frequently is enrolled into the design of everyday consumer motorcycles produced by the key MotoGP manufacturers cited above. Together with the spectacle of MotoGP racing as an event, its representation as the pinnacle of motorcycle technology is an important facet in the marketing of consumer motorcycles.

To win a MotoGP race requires the orchestration of a complex range of designed components, landscapes and networks, which are fascinating for a num-ber of reasons. MotoGP design can be seen as a process which is reworked and remade at every track through the racing season as motorcycles are tailored to individual circuits and as additional refinements are developed by manufacturers: design is never settled. Design also happens at different geographical locations: not only at different tracks themselves, but also in entanglement with the diversely located operations of parts suppliers, corporate head offices and race team head-quarters. More broadly, MotoGP provides a fascinating example of a globally mobile entity. MotoGP is a vast material event: over 80 motorcycles along with 350 tons of associated equipment must be moved from circuit to circuit (Dorna 2015). From week to week and month to month between March and November, motorcycles, tyres, fuel, spare parts, toolkits, rider motorhomes, facilities for mar-keting, hospitality and television commentary teams as well as the equipment of the dedicated MotoGP medical facility Clinica Mobile are all transported from one location to another.

Seeking to understand the design of MotoGP reveals a rich story about the multiscalar geographies of design, from the stretching of design practices across global space, to the coalescence of design understandings across and through human bodies, software systems, racing circuits, race regulations and motorcy-cles themselves. The chapter proceeds as follows: we first outline our approach to heterogeneity in MotoGP design and then work through the interconnected design elements by which teams seek to achieve racing success. We conclude by underscoring the importance of understanding the spatiality and scaling of heterogeneous design.

Heterogeneity and design

In developing an approach to understanding MotoGP design, we draw upon work in science and technology studies (STS), and particularly Law's discussions of heterogeneity (Law 2002, 2012 [1987]). Notably, Law's (2002) reflections upon the design of post-war military aircraft wings are helpful in interpreting and under-standing complex interrelations between design elements. Law's (2002) account extends through a complex range of sites, bodies and scales. Beginning with the physical forces of lift that act upon an aircraft wing, it then proceeds through such things as the embodied emotions and fears of jet pilots; the structures and preferences of governmental contracting; to the design determinants formed from the broader geographies and geopolitics of the Cold War. Opening up an extended sense of design, Law refers to the stabilising of artefact form as "heterogeneous engineering" and suggests "that the product can be seen as a network of juxtaposed components" (Law 2012 [1987], 107, emphasis in original). In an accompanying footnote he elaborates that "arguably we are all heterogeneous engineers, combining, as we do, disparate elements into the 'going concern' of our daily lives. In the present essay I am concerned, however, only with large-scale, technologically relevant system building" (Law 2012, 127). Importantly, heterogeneous engineering seeks to depict "the distribution of conditions of possibility":

> we need [...] to avoid the flattening effect of imagining that there is, on the one hand, a great designer, a heterogeneous engineer, and on the other a set of materially heterogeneous bits and pieces. Instead we need to hold onto the idea that the agent—the 'actor' of the 'actor network'—is an agent, a centre, a planner, a designer, only to the extent that matters are also decentred, unplanned, undesigned.
>
> (Law 2002, 136)

Whilst his phraseology is somewhat obtuse, we find Law's (2002) analysis valuable in thinking about the spatial stretching of design. The notion of heterogeneous engineering in Law's (2002) account seeks to address both the immediate materiality of air flow over military aircraft wings as well as more extended materialities of defence contracting and Cold War geopolitics. Further, Law (2002) seeks to foreground choices made in design—specifically attending to potential materialities which are *not* chosen as well as those which are.

Reworking Law's (2002, 2012) terminology, we refer in this chapter to *heterogeneous design* in order to capture the activity of designing. At one level our use of 'design' is intended to denote a wider reach than engineering, and is important for an analysis of MotoGP, in which engineering might imply a more narrowly technical realm. A further intent in delineating heterogeneous design is to build upon and extend conceptualisations of design from its representation as a node in commodity networks (see also Martin, Chapter 5, this volume). Through much discussion of commodity chains and networks (Leslie and Reimer 1999; Hughes and Reimer 2004) there has been an tendency to register design as one among a number of key nodes which shapes a commodity's trajectory; or to identify design as a 'moment' in the life of a commodity. Design in commodity networks has not necessarily been viewed as a discrete entity: Leslie and Reimer, for example, conceptualised furniture design "as a 'joint creation' of various actors in a broader cultural field involving multiple sites such as manufacturing, design, retailing, marketing and consumption" (2006, 322). However, the legacy of seeking to explicate different sites in the commodity chain or network may have hindered an understanding of design as a heterogeneous and multi-located set of practices.

Thus our intent in this chapter is to use the notion of heterogeneous design to foreground the web of design activity which coalesces to produce MotoGP.

There is a potential irony in Law's (2002) explication of *heterogeneous* engineering which is important to work through as we pursue our own conceptualisation of heterogeneous design. There appears a noticeable inability to capture gendered dimensions of relations among actors. For example, the gender of the (male) military aircraft pilots who appear in Law's (2002) example—even as he pursues their emotions of fear and embodied sweating—goes unremarked, leaving a generic, gender-neutral construct. In contrast we would wish to emphasise that actors are unequally positioned within design networks, based on gender differences. In the case of MotoGP, motorcycle masculinities (Pinch and Reimer 2012) are at least in part constructed through an absence of women in racing: for example, there are no women riders in the top class of motorcycle racing, although a small number of women have raced alongside men in Moto3.[2] Further, while there is much that remains unwritten about the gendering of motorsport (although see Pflugfelder 2009; Clarsen 2010) and notably motorcycle sport, it is equally important for our discussion below to be alert to gendered power relations within and across design networks (see Reimer 2016). As is evident from the narrative below, men dominate MotoGP crew chief or team manager positions as well as those of mechanical and data engineers.[3]

MotoGP: designing success

Our starting point for exploring heterogeneous design and its multiple geographies is to pose the question "what does it take to design a successful MotoGP motorcycle?". We seek to dis-assemble a range of elements necessary for victory on the race track, beginning with the designed 'object' of the MotoGP motorcycle itself. At the same time, however, we do not seek to prescribe the hierarchical importance of the motorcycle as a singular artefact. Ultimately we wish to understand the heterogeneities that shape MotoGP design.

Geographies of teams and manufacturers

MotoGP comprises a range of racing teams, from 'factory' to 'satellite' and the so-called 'Open Class' teams (Dorna 2016c). Factory teams directly represent major motorcycle manufacturers: currently Honda, Yamaha, Suzuki, Ducati and Aprilia. Austrian manufacturer KTM will join MotoGP in the 2017/18 season. Manufacturers also supply engines to satellite and open class teams, although other components in these teams' motorcycles will vary. Directly sponsored factory teams typically are the most successful. Whilst multinational motorcycle companies source components, expertise and personnel from across the globe, many of which are shared, national territorial labels still resonate in the sense that core management and engineering decisions and processes are located in the respective host countries. The headquarters of Ducati's racing team division, Ducati Corse, for example, is in Bologna in Northern Italy, and Honda's racing division Honda Racing Corporation (HRC) is located in Asaka, close to Tokyo in Japan. These centres are powerful nodes in the MotoGP network of design, not least because they are the locations from which key trajectories of engine specification and motorcycle assembly are patented and (at least initially) orchestrated and enacted. MotoGP regulations, which we discuss further below, stipulate that manufacturers must produce and then seal a certain number of engines at the start of each season, and for all manufacturers the sealed engines originate from national bases. Moreover, throughout the MotoGP season, the racetrack delivery of newly designed parts from 'the factory'—that is, from the manufacturers' national base—is a much celebrated event, not least because of the potential promise it holds for some new element of competitive advantage.

However, the production of what might first seem to be a singular MotoGP motorcycle artefact is in practice a multiplicity, with a specialist multinational character. MotoGP race engineers specify and calculate relationships between key components and subsystems, such as engines, exhausts, chassis, suspension, brakes, tyres, aerodynamics and electronics. Whilst many core components are designed and produced 'in-house', many others originate from external suppliers, such as Öhlins suspension (Sweden), Brembo brakes (Italy), Akrapovič exhausts (Slovenia), Magneti Marelli electronic control units (Italy) and Michelin tyres (France). In turn—and once again despite the national territorial labels—such subsystem components themselves comprise parts from an array of global suppliers. It should also be noted that individual race teams, and indeed MotoGP riders, are reliant financially upon a wide range of global corporate sponsors, as signified by the patchwork of brand logos and colours that adorn machines and riders.

Regulating design

The trajectories of MotoGP design are tightly prescribed by rules and regulations set down by governing body, Dorna. Detailed regulations include a wide range of engine and software specifications, covering such things as: engine types (currently four-stroke normally aspirated 1,000 cc), the materials used in construction and their modulus of elasticity[4], the permissible electronic control units (ECUs), race fuel limits, fairing and aerodynamics design, wheel and seat sizes and so on. For financial reasons—that is, in an effort to moderate team costs—MotoGP teams are only permitted to use a certain number of engines per season, which as we have indicated must be sealed at its beginning, in effect freezing further development. Other elements, however, are permitted to be modified and developed throughout the year.

Distributed design

Echoing conceptual interpretations that often characterise design as a distinctive element in commodity chains and networks, popular understandings of the role of design also tend to position it as an activity which occurs discretely, for example, prior to production or marketing. Further, design is often popularly associated with a singular and often heroic designer; a human embodiment of design 'mastery', such as a Stradivarius violin or an Alec Issigionis-designed MINI car (see also Reimer and Leslie 2008). Whilst there are parallel associations (albeit perhaps less well-known to a general public) in the recent history of MotoGP race engineering, for example Heijiro Yoshimura's 2002 five-cylinder Honda RC211V or Masao Furusawa's 90-degree crossplane crankshaft Yamaha YZF M1, the fact that a singular 'motorcycle designer' is less evident today is perhaps testimony to the distributed and collaborative practices of contemporary design.

Rather than being able to identify a single designer in a racing team—in the sense of an individual who has seen through the creation of a racing motorcycle from paper or CAD-system design to the final product—we can perhaps situate design skill across a wider range of places in MotoGP. For example, a key element of racing success can be attributed to the head of a team, whose role it is to manage dialogue across a multiplicity of technical specialisms. Interestingly, team managers or directors have a relatively wide range of educational and experiential backgrounds (although by gender they tend to be a homogenous group, as indicated above!). Some have specialist engineering training, such as Ducati Corse's Luigi 'Gigi' Dall'Igna and Honda Racing Corporation's Shuhei Nakamoto; some are former riders turned chief mechanic, such as Jeremy Burgess (the 'legendary' Australian pit crew manager for world racing champions such as Valentino Rossi, Mick Doohan and Wayne Gardner); while still others are general managers of communications and personnel, such as Yamaha's Lyn Jarvis and Honda's Livio Suppo, who have university degrees and past experience in business and management.

The motorcycle–rider assemblage

The rider is of course also central to specialist dialogues and becomes an inherent part of the co-design and engineering of the motorcycle. The practice of motorcycle design involves thinking explicitly about the 'motorcycle–rider' as an assemblage (Pinch and Reimer 2012; see also Merriman 2006 and Dant 2004 on the 'driver–car'). Design moves through the vehicle into the body of the rider; and this cyborg (Haraway 1991) is something that motorcycle designers understand and with which they work. As German designer Bernt Spiegel (2009) emphasises, the rider is "the upper half of the motorcycle". At one stage of the design process, mechanical engineers must understand the physics of motorcycle movement, such as the formalism of the equation governing maximum cornering speed limit (in short, when the centrifugal force is equal to the force of friction on the ground: see Cocco 2013, 35). However, as Law has emphasised in relation to the formalism of airplane wing lift, "the making of this centre, this formalism, performs many other relations including links that are relations of absence" (2002, 120, emphasis in original). Put another way, the 'science' of motorcycle design must deal with the fact that riding a motorcycle is highly kinaesthetic practice. Both the physiologies of MotoGP riders, in terms of their differing body weights and proportions, as well as the psychologies that shape their varied riding styles, are integral to the trajectories of design. In their search for competitive advantage, riders are adaptive to the possibilities enabled by different engine, suspension and chassis characteristics, be this 'controlled sliding', 'late breaking', or 'faster corner speeds'. Equally, however, they actively shape these characteristics as factory designers and pit crew technicians respond to these different rider styles and preferences.

MotoGP set-up: dialogues of design

The heterogeneous nature of MotoGP design also is registered through an understanding of the importance of qualifying and race 'set-up'. Set-up—which we might consider as design 'for the day'—is a process by which the racing team arrives at a pragmatic series of compromises, balancing a range of critical performance areas for the motorcycle, such as power, weight and aerodynamics, and the style and sensory feelings of its rider. And of course, skilled and careful set-up is necessary to achieve racing success. As former world champion Kenny Roberts explained:

> [R]iders, crew, and the engineers have struggled with the same basic familiar problems (and each other) for 60 years—making tires grip and last, making bikes that do what the rider wants them to, making reliable power that the rider can use.
>
> (Quoted in Cameron 2009, 6)

Central to the set-up process is an array of what might be termed co-design dialogues that occur between the rider, the pit crew (a typically multinational group of mechanical, electronic, suspension and tyre specialists working with supporting intellectual and technological infrastructures) and the motorcycle 'factory'. Fascinatingly, the 'factory' is simultaneously present at a particular racetrack through uniformed, logo-wearing mechanical and data engineering specialists in the pit crew; and absent as the domestic base described above, from where new parts may have to be supplied.

Dialogues or feedback loops between riders and set-up technicians emerge through practices connecting human and non-human actors. Factory designers and pit crew technicians must consider how the motorcycle artefact responds under varying track conditions and set-up characteristics—or what motorcyclists refer to as its 'feel'. Design information is gathered both through rider senses as well as through more quantitative methods such as data sensors; and 'the designer' becomes a diverse, distributed network. An impression of this is provided in Figure 9.1, where, filmed by the global media, Valentino Rossi can be seen providing rider feedback to his assembled pit crew (including an Öhlins suspension technician) at the 2015 Grand Prix of Qatar. On-track testing in the free practice and qualifying stages of MotoGP involves lengthy verbal discussions between riders and pit crew technicians, often exaggerated through non-verbal hand gesturing to describe key problems, such excessive 'wheelie-ing', front end 'chatter', lack of rear wheel grip

Figure 9.1 Valentino Rossi and the pit crew, Losail International Circuit, Qatar, 2015.

and cornering instability. Not all riders have extensive knowledge of the technical vocabularies of motorcycle engineering and so the ability to translate the machine language of the rider is an important requirement for pit crew staff. Some teams also deploy spotters stationed at key points at the side of the racetrack to observe the performance of the rider/motorcycle assemblage.

Non-human feedback is derived in part from automated trackside cameras filming rider performance, which simultaneously stream images to television audiences and pit crews. Further, the many data sensors, accelerometers and gyroscopes attached to the motorcycle provide an extensive array of performance statistics. Post-event analysis of serious material failures, such as engine seizures and blow-ups, tyre de-laminations or ECU malfunctions, provides important feedback to MotoGP teams and riders. For riders, the most dramatic form of feedback occurs at the point of a crash, when they are said to have "found the limit" of mechanical, electronic and bodily capacities. The moment of a crash is often described by riders as the point at which the motorcycle "lets go". At this moment the motorcycle–rider assemblage literally falls apart—is disassembled—with potentially perilous consequences for the rider, although even these impacts are mediated by the design of increasingly sophisticated crash helmets and air-bag technology leather race suits.

Design intermediaries: tyres, technologies and tracks

The MotoGP race weekend practice, qualifying and warm-up sessions are key spaces and times where these set-ups are designed, not least because of the significance of the different geographies and layout characteristics of each track. The sessions, however, also build upon the inputs of factory test-riders whose publicly unseen work at a range of locations continues throughout the year. The engineering science of set-up means that the slightest of changes to the motorcycle artefact, such as the micro-millimetre positioning of and relationships between engines, wheels and forks, can have dramatic effects upon motorcycle handling and stability (Spalding 2010). Equally, however, pit crews and managers know that they are designing a rider–motorcycle assemblage; and a crucial part of their work involves reassuring and building the confidence of the rider. Cameron (2009, 162) makes this point by quoting Öhlins' suspension technician Jon Cornwell: "Most of what we do here (at the track) isn't engineering. It's making riders comfortable, giving them confidence". The popular adage amongst pit crews that 'it is much less expensive to fix the rider than the machine' is testament to a more extended and embodied sense of design.

The corporeal and embodied nature of MotoGP racing is arguably most vivid through the curves and corners of a racetrack, where riders can achieve lean angles in excess of 65 degrees, presenting a testimony to their skill and bravery (see Figure 9.2). Equally, however, the preceding discussion also has illustrated that the motorcycle–rider assemblage is mediated and designed through a complex combination of human and non-human infrastructures. Complex electronic rider aids, such as traction control systems, have become increasingly important over recent years. These non-human or quasi-human 'delegates' of 'actants' (see Latour, 1992, 1993), which have now filtered into consumer motorcycles, continually intercede in the millisecond processes and governances of movement (Pinch and Reimer 2012). In MotoGP, where technology is most developed, simultaneous wireless communication with electronic transponders positioned around the circuit are used to modify the set-up and power characteristics of the motorcycle–rider assemblage, reworking rider inputs such as accelerating, braking and bodily positioning, so that the optimum performance is produced for individual corners and sections of the racetrack. In this sense, design is never complete, but subject to continual adaptation and modification as riders circulate around a given track.

Illuminating the role and nature of such technologies provides further insight to the multiplicities and complexities of design and the networked geographies and locations from where it is produced. This point can be extended further through a consideration of another actor: the motorcycle tyre. Although tyres are crucial to performance in all forms of motor sport, they are much more significant in motorcycle racing because of the very narrow contact patch between rubber and surface, compared, for example, with the wide four wheels of a Formula 1 racing car. Whereas historically teams were allowed to source tyres from competing manufacturers, they are now restricted by Dorna regulations to those from a single supplier. Michelin, the current designated tyre manufacturer, supply a range of tyre options of different compounds, and for dry and wet weather conditions, to the race teams.

Much is made of the impacts of tyres upon motorcycle set-up, particularly in relation to what riders describe as the 'feel' the tyres generate. Moreover, the material qualities of different tyres impact upon the styles riders adapt and

Figure 9.2 MotoGP riders Marc Marquez and Bradley Smith at the Laguna Seca Raceway, California, USA, 2013.

develop (Cameron 2014). When Bridgestone was the designated manufacturer a renowned characteristic was the ability of the front tyre to absorb extreme 'load' on entry into corners, a fact exploited by Honda factory rider Marc Marquez during two championship-winning seasons. Alongside personal tyre preferences based on such embodied 'feel', different tyres are said to favour the structures and engine characteristics of some motorcycles over others, to the extent that they drive the design trajectories of engine and chassis configurations. Equally, however, tyre design is driven by a multiplicity of other conditions and factors. Each MotoGP race circuit has a range of different geographies and characteristics, be this more or less surface abrasion, longer straits, more left or more right-handed corners and so on. These elements are never static, for example, track and air temperatures and weather conditions change through a race weekend, or circuits are reconfigured and resurfaced over time. Consequently, a range of bespoke tyres are made for each circuit. As much as MotoGP success is designed through extraordinary and complex human and non-human knowledges, technologies and infrastructures, it is also a product of the mundane geographies of sunshine and rain, asphalt and gravel.

Conclusion

The chapter has sought to understand MotoGP as heterogeneous design: a distributed, multifaceted and spatially extensive entity. Whilst this elite racing class may seem an exceptional example, we would argue that its connection with more quotidian forms of motomobility illuminates design as networked and distributed. The extended realm of heterogeneous design is signalled in part by the ways in which technologies developed for MotoGP increasingly feature in consumer motorcycles, helmets and clothing. We have emphasised the complex set of non-additive relationships in MotoGP design, which connect motorcycle components (chassis, tyres, engine, brake systems and so on); rider skill and experience; electronic control systems and software; the configuration of race circuits and the skill of team managers. A key insight is that successful design involves the creation of an artefact that is flexible enough to accommodate the need for constant redesign. Thus whilst an individual or team responsible for digital or workshop prototype design might still be considered to be 'the designer', they also must materialise the distributed nature of design in the artefact itself by building flexibility and in working through various feedback mechanisms.

We have stressed that design is never complete: MotoGP racing design is differently performed and achieved at each circuit, from day to day and race to race, through set-up changes for technical reasons, as well as in response to rider feedback and momentary changes in environmental conditions. MotoGP design also can be seen as unfinished and incomplete because of the ongoing role of testing. At the end of the race weekend at certain circuits, riders begin to test possible design options to be implemented not only for the current, but also for the following season.

The chapter has utilised the notion of heterogeneous design in order to understand design as a multi-located set of practices. The difficulty of achieving racing success—and particularly repeated success over the MotoGP season—bears out Law's (2002, 136) characterisation of heterogeneity as a "process full of tension", in which the "conditions of possibility" are "both present and not present". Our contribution is then to spatialise this imagination and to delineate the geographies of heterogeneous design. We have done this through an exploration of the sites and locations through which MotoGP design works. These range from the riding style of an individual competitor; through the rider's locally performed relationships with a pit crew; the particular attributes of individual racetracks; as well as global networks of motorcycle manufacturing and component supply. In exploring heterogeneous design by moving from the motorcycle–rider assemblage to a broader, multi-located set of practices we enliven our understanding of geographies of design.

Notes

1 At the time of writing, Grand Prix racing is split into three classes: Moto2, Moto3 and the 'top class' of MotoGP (the latter with a maximum permitted engine displacement of 1000cc). Because Moto2 and Moto3 are run under differing technical and rider regulations, our discussion in the chapter is restricted to the main class.

2 Spanish riders Ana Carasco and Maria Herrera have both competed in Moto3; and notable women in other categories of motorcycle racing include British Superbike rider Jenny Tinmouth (UK) and Spanish rider Laia Sanz, who has competed alongside men in the Dakar multi-country off-road rally.

3 While there has been some interesting work on the birth of the Women's Engineering Society (Pursell 1993; Phipps 2002), considerably more research is needed to unpack historic codings of engineering and technical skill as essentially masculine.

4 This is a measure of stiffness of an elastic material: it is used to describe its properties when stretched or compressed.

References

Cameron, K. 2009. *The Grand Prix motorcycle: the official technical history.* Yeovil, Somerset: Haynes Publishing.

Cameron, K. 2014. MotoGP: origins of style. 6 August 2014. Available at: http://www.cycleworld.com/2014/08/06/motogp-racing-evolution-of-motorcycle-riding-styles-part-three-of-three-by-kevin-cameron, accessed 27/06/16.

Clarsen, G. 2010. Automobiles and Australian modernisation: the Redex Around-Australia trials of the 1950s. *Australian Historical Studies* 41: 352–368.

Cocco, G. 2013. *Motorcycle design and technology.* Vimodrone (MI), Italy: Giorgio Nada Editore.

Dant, T. 2004. The driver-car. *Theory, Culture, and Society* 21: 61–79.

Dorna Sports Ltd. 2015. MotoGP signs official logistics partner deal with DHL. Available at http://www.motogp.com/en/news/2015/07/17/motogp-signs-official-logistics-partner-deal-with-dhl/180240, accessed 27/06/16.

Dorna Sports Ltd. 2016a. Inside MotoGP: Overview. Available at: http://www.motogp.com/en/Inside+MotoGP/Overview, accessed 21/06/16.

Dorna Sports Ltd. 2016b. Dorna: the company. Available at: http://www.dorna.com/dorna_thecompany.html, accessed 21/06/16.

Dorna Sports Ltd. 2016c. Bikes. Available at: http://www.motogp.com/en/Inside+MotoGP/Bikes, accessed 27/06/16.

Haraway, D. 1991. *Simians, cyborgs, and women: the reinvention of nature.* London: Free Association Books.

Hughes, A. and Reimer, S., Eds. 2004. *Geographies of commodity chains.* London: Routledge.

Latour, B. 1992. Where are the missing masses? The sociology of a few mundane artefacts. In *Shaping technology/building society: studies in sociotechnical change.* Ed. W. E. Bijker and J. Law. Cambridge, MA: MIT Press, pp. 225–258.

Latour, B. 1993. *We have never been modern.* Cambridge, MA: Harvard University Press.

Law, J. 2002. On hidden heterogeneities: complexity, formalism, and aircraft design In *Complexities: social studies of knowledge practices.* Ed. J. Law and A. Mol. Durham, NC: Duke University Press, pp. 116–141.

Law, J. 2012 [1987]. Technology and heterogeneous engineering: the case of Portuguese expansion. In *The social construction of technological systems: new directions in the sociology and history of technology.* Ed. W. E. Bijker, T. P. Hughes and T. Pinch. Cambridge, MA: MIT Press, pp. 105–127. Anniversary edition. [Originally published 1987.]

Leslie, D. and Reimer, S. 1999. Spatializing commodity chains. *Progress in Human Geography* 23: 401–420.

Leslie, D. and Reimer, S. 2006. Situating design in the Canadian household furniture industry. *The Canadian Geographer* 50: 319–341.

Merriman, P. 2006. 'Mirror, signal, manoeuvre': assembling and governing the motorway driver in late 1950s Britain. *The Sociological Review* 54: 75–92.

Pflugfelder, E. H. 2009. Something less than a driver: toward an understanding of gendered bodies in motorsport. *Journal of Sport and Social Issues* 33: 411–426.

Phipps, A. 2002. Engineering women: the gendering of professional identities. *International Journal of Engineering Education* 18: 409–414.

Pinch, P. and Reimer, S. 2012. Moto-mobilities: geographies of the motorcycle and motorcyclists. *Mobilities* 7: 439–457.

Pursell, C. 1993. 'Am I a lady or an engineer?' The origins of the Women's Engineering Society in Britain, 1918–1940. *Technology and Culture* 34: 78–97.

Reimer, S. 2016. 'It's just a very male industry': gender and work in UK design agencies. *Gender, Place and Culture* 23: 1033–1046.

Reimer, S. and Leslie, D. 2008. Design, national imaginaries, and the home furnishings commodity chain. *Growth and Change* 39: 144–171.

Spalding, N. 2010. *MotoGP technology,* second edition. Yeovil, Somerset: Haynes Publishing.

Spiegel, B. 2009. *Upper half of the motorcycle: on the unity of rider and machine.* Trans. M. Hassall. Conway, NH: Whitehorse Press.

10 Universalising and particularising design with Professor Kawauchi

Kim Kullman

Introduction

On returning to Tokyo from a recent trip to London, Yoshihiko Kawauchi, an architect and professor, retreats to his office and begins to sift through the materials that he has gathered during his week-long journey (Figure 10.1). He first uploads all the photos onto his computer and studies them, using the images to evoke his experiences of the buildings and spaces that he encountered on site visits. Then Kawauchi listens to interview recordings he made with British design experts, selecting useful ones for transcription and translation into Japanese. Finally, he looks online for relevant regulations, guidelines and standards and downloads them into separate, meticulously arranged folders on his desktop. This is a simple but effective technique that Kawauchi employs to compare different

Figure 10.1 Kawauchi working at his desk.

national approaches to design practice, bringing together places from around the world on his computer screen.

At the time of writing, the empirical materials that Kawauchi brought back to Tokyo have gradually percolated into different areas of the Japanese society. Kawauchi, for example, has given a talk to 300 policymakers and politicians in the parliament, where he discussed the forthcoming Tokyo 2020 Olympics, drawing on his tour of diverse sporting facilities in London. Likewise, Kawauchi has appeared on national radio and in newspapers, again discussing the accessible design of the Olympics—a topic that at the time of writing attracts wide interest in Japan. Apart from circulating his experiences of London through Power Point slides, radio programmes and articles, Kawauchi travels extensively, including to countries collaborating with a Japanese development organisation that often invites him to speak to local professionals and students. Thus Kawauchi exports his ideas out of Japan again.

* * * * *

This chapter contributes to the emerging field of *design mobilities*, which explores the varied ways whereby ideas, models and techniques of design are developed, disseminated and transformed as they travel around the world, whether in the guise of types (Guggenhem and Söderström 2010), patents (de Laet 2000), standards (Busch 2012), machine parts (Morita 2013), raw materials (Knowles 2013), methods (Gunn *et al.* 2013), policies (McCann and Ward 2011) or experts (Hult 2015). A common question across this work is the extent to which design mobilities shape and become shaped by transnational relations of knowledge and power as well as how such relations eventually influence local built environments. Research indicates that these exchanges of design ideas and models have only intensified of late due to the liberalisation of markets, the global interconnectedness of organisations and the increasing mobility of professionals (Guggenheim and Söderström 2010, 8–14).

The protagonist of this chapter, Professor Yoshihiko Kawauchi, is variously implicated in contemporary design mobilities. For several decades he has been developing a concept known as universal design, which seeks to produce objects, spaces and services that are accessible for the widest range of people, including, but not limited to, elderly and disabled persons. The most common definition of the concept can be found in the UN Convention on the Rights of Persons with Disabilities, adopted in 2006, where it denotes the "design of products, environments, programmes and services to be usable by all people, to the greatest extent possible, without the need for adaptation or specialized design" (UN 2006). Often this general formulation is further clarified by reference to the "seven principles" elaborated by a group of American academics and experts in 1997: "equitable use", "flexibility in use", "simple and intuitive use", "perceptible information", "tolerance for error", "low physical effort" and "size and space for approach and use" (CED 2006). Combined with an array of resources, including standards, best practice models, case studies, professional training courses, conference presentations and websites, these attempts to formalise universal design have created a widespread narrative emphasising the seamless travel of the concept and the relative ease with which it can be implemented across different environments.

As this chapter argues, however, a close engagement with the practices of Kawauchi reveals a universal design that is more unsettled than assumed and consequently open to debate over its content and shape. In doing so, the chapter develops understandings of design mobilities in two related ways. First, it explores a topic that has received little attention to date: the role of personal, embodied mobility in the transmission of design concepts. Many studies on the efforts of architects, planners and other "members of epistemic, expert, and practice communities" (Peck and Theodore 2010, 170) to make design travel tend to reduce people to "the archetypal cosmopolitan" (Guggenheim and Söderström 2010, 12), who glides from one place to another, effortlessly adopting, disseminating and discarding ideas along the way. Yet, when turning to Kawauchi, we are reminded that persons, and particularly their embodied experiences, can occupy a central position in the mobilisation of design concepts. A person-oriented approach does not have to entail subscribing to a narrowly humanistic mode of thinking, but rather highlights individuals as distributed entities that bring together diverse sites, materials and practices. Building on anthropological theory (Mosko and Damon 2005; Strathern 2004; Wagner 1991) the chapter will therefore argue that Kawauchi is a "fractal person", who mediates between the global flows of ideas and materials that make up universal design, and thereby comes to act as a source of influence in his field.

Second, the chapter addresses the geographical complexity of universal design by demonstrating that tracing the mobility of this concept "requires exploring more than simply the diffusion or circulation of a self-contained policy or planning model; it entails examining the milieu or factual terrain through and in which it has been shaped" (Harris and Moore 2013, 1503). Universal design cannot be taken as a ready-made entity that experts may conveniently detach from particular environments and transplant into other settings (for criticism, see Hamraie 2013; Imrie 2012; Lid 2013). Instead, it emerges in a precarious process of *universalisation* (Jullien 2014, 100–120), whose main purpose is to assemble and circulate the concept, so that its global applicability and influence increases. Nonetheless, this endeavour also implicates a concurrent tendency of *particularisation* (Imrie 2012, 879–880), where ideas, materials and sites of universal design reveal themselves to be embedded within specific bodies and spaces, which travel only with difficulty and complicate attempts to generalise across corporeal, cultural and geographical differences. The chapter does not argue against universality in favour of particularity, because this would overlook that universals form "bridges, roads, and channels of circulation" (Tsing 2005, 7) that facilitate connection, communication and knowledge exchange. It does, however, suggest that critical attention must be directed to specific practices that enable the dissemination of universal design, as these may mask other understandings of the concept. This will lead to an exploration of alternative ways of negotiating the universals and particulars of current design mobilities.

The subsequent discussion builds on a case study for a European Research Council funded project (number 323777), which involved extensive fieldwork with Kawauchi, including in-depth interviews and participant observations during site visits and meetings in London and Tokyo. Throughout, the aim was to facilitate an interdisciplinary dialogue on universal design in order to explore how the concept is presently developed and how it might evolve in the future. Collaborating closely with Kawauchi, the study demonstrates that researchers of design mobilities are as entangled with such mobilities as the people they study: Kawauchi not only shared our interest in uncovering the complexities of universal design, but also worked in a manner similar to ours, deploying a range of methods, from observations to interviews and photography, to trace a concept on the move.

Kawauchi as a fractal person

The understanding of Kawauchi as a fractal person is indebted to a strand of anthropological thinking that draws inspiration from mathematics and complexity theory to bring new conceptual resources into the study of topics such as scale, comparison and personhood (see Green 2005; Holbraad and Pedersen 2009; Mosko 2005; Strathern 2004; Wagner 1991; 2001). According to the mathematical definition, a fractal "refers to phenomena of 'self-similarity', or the tendency of patterns or structures to recur on multiple levels or scales" (Mosko 2005, 24). As an example, we may take the fern, a plant that displays a recurrent triangular pattern at a range of scales, from the stem to the leaflets. Another common example is any coastline, which retains its irregular and jagged pattern irrespective of whether it is viewed from afar or up close (Mosko 2005, 24). What both of these examples share is self-similarity, as they keep their complexity across scales, which is "reproduced regardless of the detail on which one zooms in" (Jensen 2007, 832).

Among the first to experiment with the possibilities of fractality was Wagner (1991), who introduced the concept of "fractal person" to describe Melanesian personhood as expressed in relations of kinship, power and wealth. For Wagner, fractality enables an exploration of the complex ways in which Melanesians conceive of the person as an open-ended, relational entity that is neither fully an individual nor a group, neither singular nor plural, neither whole nor part. Instead, a "fractal person is never a unit standing in relation to an aggregate, or an aggregate standing in relation to a unit, but always an entity with relationship integrally implied" (Wagner 1991, 163). Wagner argues that persons—who are constituted in relations, and can therefore only be understood through those relations—carry complexity within, much like the fern and the coastline discussed above. This means that a single person can bring together bodies, materials and spaces from near and far in ways that problematise simple distinctions between the local and the global, the micro and the macro. As Jensen (2007, 844) argues,

A fractal approach exempts the scholar from scaling contexts, people, and places as to their size and importance, but enables the observation of how people are themselves constantly instantiating scales and hierarchies through building and maintaining relationships with other people, with institutions, and with objects.

For the present argument, fractal thinking offers a way into the practices of Kawauchi, who could be seen as forming an intricate arrangement of human and nonhuman materials that he mobilises to shape universal design. Kawauchi brings universal design into being by turning himself into a "field of relations" (Strathern 2009, 150) that spans cultural and geographical distances. At the same time, he exposes universal design as an emergent entity, "generated by the interrelations between things, rather than by the essential characteristics of things themselves" (Green 2005, 136). This has a bearing on current work on design mobilities, which demonstrates that scaling, comparing and measuring are central practices in the global transfer of ideas, enabling experts to establish relations of similarity and difference across dispersed sites (Guggenhem and Söderström 2009; McCann and Ward 2011; Morita 2013). Such practices are often seen as simplistic in that they overlook the situatedness of design by reducing it to models, numbers and other transportable entities (Ward and McCann 2011, 175–177). Kawauchi, however, indicates that it is near impossible to find a vantage point from which to stabilise and align sites so as to scale, compare or measure them. As Strathern (2004, 24) writes about cross-cultural comparison, it is usually the case that "there is no proportion between [places], no encompassing scale or common context that will make these […] units of a comparable order". Indeed, as will become clear below, the process whereby Kawauchi composes and disseminates universal design is a highly precarious one.

Embodying universal design

Let us now return to where we started, the desk of Kawauchi, which is a place where he sets into motion various types of design ideas, many of them explored in ユニバーサル・デザイン—バリアフリーへの問いかけ, a book that was originally published in 2001. Currently in its seventh edition in Japan, the book has been translated into Korean and English, appearing in the latter as *Universal Design: A Reconsideration of Barrier-Free* (2009). Travelling around the world, it has become an "extension of the author's person" (Demian 2004, 70), contributing to Kawauchi's influence over the field of universal design by leading to invitations to give talks at conferences and take part in projects. Written for professionals in an accessible language, the book is concerned with the concept of universal design and how its "expansion […] has the potential to change the nature of the concept itself' (Kawauchi, 2009, vii). It recounts numerous research trips made by Kawauchi to the United States with the aim of comparing, through site visits and interviews with experts, local universal design standards, best practice examples, regulations and laws with those in Japan. Apart from engaging in comparison, the text adopts a distinctively autographical tone, as Kawauchi draws upon personal experiences to elaborate his argument. In one evocative passage, he describes being helped onto a complicated staircase lift by the staff at a train station, while other passengers are watching intently. Kawauchi thereby shows that specialised design solutions direct unwanted attention to people by, as he calls it, "emphasising disability" (Kawauchi, 2009, 17)—the lift cannot be universal design, because it does not place all persons, regardless of their abilities, on equal footing.

This account is compelling because Kawauchi relates personal, embodied experiences to broader issues faced by disabled people. Kawauchi became a wheelchair user after sustaining a spinal cord injury in a sporting accident in the early 1970s, and his views of design are shaped by having trained as an architect when accessibility was rarely considered a topic worthy of attention among Japanese design professionals. As he repeatedly emphasises, "my specialty is being a wheelchair user," meaning that mobility forms the basis of his design thinking and allows him to cultivate a "view from a body rather than [a] view from above" (Strathern 2004, 32) (Figure 10.2). Arguing that "drawings or documents, [...] do not give me the reality", Kawauchi prefers to experience design first hand to explore whether solutions praised in the specialist literature work in everyday settings. For him, abstract measurements—that a ramp has "a 1:12 steepness"—cannot convey a "physical impression" of universal design.

Although working through bodily experiences may differ from other instruments for evaluating design, it serves various purposes. Kawauchi, for example, transforms himself into a site that brings culturally and geographically dispersed entities into specific kinds of relations, based on his capability to shift scales from local to global, particular to universal, "from small to large and, possibly, back to small

Figure 10.2 Kawauchi in the Tokyo transport system.

again" (see Jensen 2007, 846). He illustrates how select individuals around the world—many of whom are included in his book—act as relays for the international flows of ideas, models and materials that constitute universal design, becoming fractal persons with "relationship integrally implied" (Wagner 1991, 163). The remainder of this chapter therefore describes how Kawauchi, as a mediator between the travelling entities that make up universal design, engages in the "universalization of particularism and the particularization of universalism" (Robertson 1992, 102). He is both *universalising* design, or combining particulars into universals, and *particularising* design, or decomposing universals into particulars.

Universalising particulars

Kawauchi offers us insight into how universal design is assembled from particulars, gathered during tours abroad and incorporated into slides, reports and other materials in an attempt to imbue them with qualities and values of the best practice model—a process that generally entails "detaching given features from existing, locally rooted" design in order "to make them reproducible" (Guggenheim and Söderström 2010, 5–6). For Kawauchi, universal design is not a definite object that exists independently of everyday situations, but rather an outcome of a continuous process of *universalising* that "does not claim but creates, and its value is measured by the power and intensity of this effect" (Jullien 2014, 117).

To develop this argument, we need to visit Haneda, one of the two major airports in Tokyo. Located a 30-minute train ride from Tokyo Station, Haneda serves mostly domestic passengers, although it has steadily increased the number of international flights, currently handling some 28 per cent of foreign passengers arriving in Tokyo annually (Fujikawa, 2014). In 2010, Kawauchi acted as the chair of the access panel developing a universal design plan for the new international terminal. He participated in the selection of the panel members and, after the completion of the building, became involved in the promotion of the airport, for example giving international presentations on Haneda, including a talk in London, where he described how the panel worked to "evaluate the effectiveness" of universal design.

Kawauchi is not the only actor seeking to universalise Haneda by turning it into a circulating entity. The airport has been lauded widely in media outside Japan (see Independent 2011; Sensalis 2014), and adding to its global exposure are the websites of Haneda Airport (2016) and Japan Airlines (JAL 2016), which contain multi-lingual information on the universal design features of the new terminal, including maps and images. The most revealing aspect for the present argument is the way Haneda brings out the embodied character of Kawauchi's universalisation practices, an illustrative example being the visits that he arranges for design experts (Figure 10.3). Similar activities have been discussed in recent studies on design mobilities, where professionals are described as taking part in tours where "the places they visit are carefully regulated by their organisers" (Temenos and McCann 2013, 349), often resulting in encounters that are divorced from the complexities of everyday life (see McCann and Ward

Figure 10.3 Kawauchi and his associates at Haneda.

2011; McFarlane 2011; Peck and Theodore 2010; Peck and Theodore 2012). Haneda is certainly conducive to such controlled tours by being easily accessible from Tokyo and by providing a selection of universal design that can be viewed in a relatively short space of time inside an air-conditioned building that is dotted with cafes and restaurants.

During the tour that Kawauchi organised for our team, we followed a scripted itinerary that included most of the items listed on the Haneda Airport website (2016). Nevertheless, the more closely we engaged with the specificities of the airport, the more complexity we encountered. Kawauchi arranged for a group of professionals, many of them members of the access panel, to guide us through the terminal and share their opinions, including two airport officials, a sign designer and a visually impaired expert. Moving as part of this collective, it was apparent that Kawauchi and his associates were collaborating to perform a certain version of universal design for us. We were also drawn into this performance by being asked to stop and touch materials, listen to sounds and consider how diverse types of bodies, from a visually impaired person to a wheelchair user, might respond to the airport. All along, Kawauchi was demonstrating how he, as a wheelchair user, could operate with ease in various spaces, such as toilets, where he would open and close doors, turn around and reach for buttons and taps.

The tour affirmed the status of Kawauchi as a fractal person, who has the ability to gather actors and involve these in a choreographed enactment of universal design. However, it also illustrated that universal design is not a ready-made and stable concept but grounded in specific bodies and spaces. As Laclau (1996, 35) argues, universality builds on particularity, because it is based on competing attempts by people "to temporarily give to their particularisms a function of universal representation" (see Imrie 2012, 879–880). The tour of Haneda, then, did not imply a definitive notion of universal design, but came across as a set of performative practices that are "on the move" and "not yet concluded" (Jullien 2014, 116). Despite the efforts to enact a coherent version of universal design, we became embroiled in an endless series of particulars, from buttons and door handles to font types and announcements. Bringing our bodies—shaped by differing capabilities, experiences and assumptions—to bear on Haneda, we exchanged impressions of particulars, debating and sometimes disagreeing over the extent to which these could include diversity and difference. It is to this particularisation of the universal—"inviting it to retreat from its global imperialism and [...] putting it back in its proper place" (Jullien 2014, 153)—to which we now turn.

Particularising universals

Kawauchi is fully aware of the challenges of universalisation, a case in point being the very practice of everyday travel, which is central to his dissemination activities. As a wheelchair user, he must plan his journeys meticulously, although most trips still involve unexpected obstacles—for example, during a visit to London, he was forced to take a taxi to a stadium promoting itself as universally designed, because it could not be reached in a wheelchair by public transport. These often frustrating moments particularise universal design for Kawauchi, or expose its situatedness in "the sticky materiality of practical encounters" (Tsing 2005, 1). Significantly, such particulars are regarded by Kawauchi as containing potentially the same complexity as the seemingly universal, global or large (Strathern 2004, xix), and may sometimes even bring into question whole universal design projects.

Let us consider another place that Kawauchi toured during his journey to London: Queen Elizabeth Park, the 560-acre site for the 2012 Olympics. The visit was one of several that he conducted during a week-long trip, which served as a fact-finding mission for the Olympic Committee of Koto, a ward of Tokyo, where half of the new sports venues will be constructed for the 2020 Games. As research has amply demonstrated, the Olympics are often constituted as a specific kind of knowledge object by experts and professionals around the world, from events managers to designers, who often approach it as a set of best practice models that can be disconnected from their actual surroundings and used successfully in other places (see Hayes and Horne 2011; Rowe 2012; Watt 2013). One example of this are the universal design guidelines created for the Olympics and now distributed worldwide as a template for future games and other major sports events (LLDC 2013). Most design experts in London mentioned this document, advising Kawauchi to download and study it carefully. In the guidelines, the Olympics are performed as an undisputable success story, catering for a variety of users: "older people, people with temporary impairments, large families, parents with young children and babies, people from diverse faith groups and different cultures, people that speak different languages and a combination of all of the above" (LLDC, 5).

However, as Kawauchi toured the premises, he learned that the venues were too far apart from one another, making the distances between them quite long for him to manage in a manual wheelchair. And inside the Velodrome, several spaces reserved for spectators in wheelchairs had been eliminated through the addition of regular seating. Likewise, as we moved into the Olympic Village—or East Village, as it has been subsequently renamed—there were no kerb cuts in sight on an otherwise calm street that allows pedestrians ease of movement, making it difficult for Kawauchi to cross the road. Finally, at the Lee Valley Hockey and

Figure 10.4 Kawauchi on a visit to Queen Elizabeth Park.

Tennis Centre, which hosted wheelchair tennis during London 2012, Kawauchi visited the changing facilities and noticed that many toilets had features to assist disabled people, but few could accommodate tennis players, who have wider wheelchairs (Figure 10.4).

Kawauchi, then, seemed to scale down the grandeur of the universal design of Queen Elizabeth Park, turning it into a set of specificities: toilet door widths, kerb cuts, seating for wheelchair users. These specificities did not always cohere as expected, hence counteracting the narrative of the guidelines, which gave an impression of a uniformly applied design approach and a universally accessible environment. Kawauchi's embodied critique clearly indicated that the degree to which universal design works cannot be evaluated once and for all within the distanced, ordered and quantitative space of guidelines, but only on a continuous, everyday basis by the countless individuals who actually engage with built form.

During his journeys, Kawauchi often finds himself in similar situations, where "parts continually proliferate more parts" (Green 2005, 137), as he encounters an endless series of new design features, which he then tries to record and later organise and analyse at his desk in Tokyo. However, particulars seldom combine seamlessly and have a tendency to be become excluded from Power Points, articles, books and reports. For example, on the slides that Kawauchi made for a talk at the Japanese parliament after the London trip, the above moments are not mentioned, despite their significance at the time. This suggests that many of the productive encounters experienced and recorded by travelling design experts might not find their way into talks, publications, guidelines and, further down the line, into built environments. This raises the question of exactly what function particulars can have in universal design and whether experts could enrich the concept through becoming more receptive to the irreconcilability of the materials and knowledges that constitute it.

Certainly for Kawauchi, as particulars continue to pile up from one journey to another, universal design is becoming an increasingly fragile concept to hold together. Illustrative of this is the case of Jordan, a country that Kawauchi has visited as an ambassador for the Japan International Cooperation Agency (JICA). JICA guidelines underline the importance of "introducing universal designs when facilities/equipment are designed and constructed with JICA's assistance" (2009, 7).

For Kawauchi, however, exporting Haneda, the Olympics, or any other globally circulating universal design example to Jordan seems challenging in a place where other problems are more pressing. It has become apparent to him that the entities mobilised by experts may not be replicable universals, but particulars that carry cultural and geographical specificity and travel only with difficulty.

Conclusion: cross-cultural compatibility

Taken together, the above discussion highlights two specific challenges for future work on universal design. First, it points to an alternative design concept that is based on what Tsing (2005, 8) calls "engaged universals", which "travel across difference" but are "charged and changed by their travels." Concentrating on the mobility practices of Kawauchi, the chapter has demonstrated that universal design does not only move as reports, best practice models and other downloadable entities, but *within* and *as* bodies. Although a recognition that universal design is sustained in bodily practices serves as a reminder of its situatedness and also the inescapable difficulties of scaling, measuring and standardising the concept, few global design experts seem to cultivate the kinds of grounded knowledges described here. One way forward might therefore be to acknowledge that single individuals, organisations and sites cannot be taken as models for universal design, because, in its embeddedness, the concept is empirically specific. This invites caution about the kinds of generalisations that professionals make about the world and the role of design in it, as "engaged universals are never fully successful in being everywhere the same" (Tsing 2005, 10).

Second, although it may be difficult to establish equivalences among the particulars of universal design, it is possible to practise a form of cross-cultural "compatibility" that seeks to make "connections without assumptions about comparability" (Strathern 2004, 38). In the case of Jordan, for example, Kawauchi struggled in his wheelchair on the pavements, but was struck by the willingness of people to help him on the street, thus compensating for the lacking infrastructure. As Kawauchi explained, the Japanese situation is often the reverse, because infrastructures work, but people are less inclined to help. This has made him realise that universal design is overly concentrated on technical solutions, while failing to foster interpersonal engagement and care. Such, ostensibly modest, cross-cultural encounters demonstrate that even if the attempt to universalise design seems to generate more and more corporeal, cultural and geographical particulars, certain relations of compatibility can still be established across differences—relations that avoid subsuming the entities they bring together into one model of universal design, allowing them instead to enrich and transform the notion as it travels from one place to another. Indeed, as the work of Kawauchi suggests, universal design is not a ready-made concept, but an open arrangement of ideas, materials and practices, whose purpose is not necessarily to become fixed so that it can be evaluated, measured or compared by experts, but rather to facilitate an evolving, admittedly challenging, debate around the possibility of societies to create environments for the widest range of capabilities.

References

Busch, L. 2012. *Standards: recipes for reality*. Cambridge, MA: The MIT Press.

Center for Universal Design (CED). 2006. *The principles of universal design*. North Carolina State University Center for Univeral Design. Available at https://www.ncsu. edu/ncsu/design/cud/pubs_p/docs/poster.pdf, accessed 3/2/16.

de Laet, M. 2000. Patents, travel, space: ethnographic encounters with objects in transit. *Environment and Planning D: Society and Space*, 18: 149–168.

Demian, M. 2004. Seeing, knowing, owning: property claims as revelatory acts. In *Transactions and Creations. Property Debates and the Stimulus of Melenesia*. Ed. E. Hirsch and M. Strathern. London: Berghahn Books, pp. 60–82.

Fujikawa, M. 2014. Tokyo's Haneda Airport challenges Narita. *The Wall Street Journal,* 18 December. Available at http://www.wsj.com/articles/tokyos-haneda-airport-challenges-narita-for-international-passengers-1418959342, accessed 03/03/16.

Green, S. 2005. *Notes from the Balkans: locating marginality and ambiguity on the Greek-Albanian border*. Princeton, NJ and Oxford: Princeton University Press.

Guggenheim, M. and Söderström, O. (Eds) 2010. *Re-shaping cities. How global mobility transforms architecture and urban form*. London: Routledge.

Gunn, W., Otto, T. and Smith, R. C. (Eds) 2013. *Design anthropology: theory and practice*. London: Bloomsbury Academic.

Hamraie, A. 2013. Designing collective access: a feminist disability theory of universal design. *Disability Studies Quarterly*, 33. Available at http://dsq-sds.org/article/view/3871, accessed 3/2/16.

Haneda Airport. 2016. *Barrier free information*. Available at http://www.haneda-airport. jp/inter/en/premises/service/assist.html, accessed 20/11/16.

Harris, A. and Moore, S. 2013. Planning histories and practices of circulating urban knowledge. *International Journal of Urban and Regional Research*, 37: 1499–1509.

Hayes, G. and Horne, J. 2011. Sustainable development, shock and awe? London 2012 and civil society. *Sociology*, 45: 749–764.

Holbraad, M. and Pedersen, M. A. 2009. Planet M: the intense abstraction of Marilyn Strathern. *Anthropological Theory*, 9: 371–394.

Hult A. 2015. The circulation of Swedish urban sustainability practices: to China and back. *Environment and Planning A*, 47: 537–553.

Imrie, R. 2012. Universalism, universal design and equitable access to the built environment. *Disability and Rehabilitation*, 34: 873–882.

Independent. 2011. Haneda sets standards for barrier-free airports. *The Independent*, 23 April.

JAL. 2016. *JAL universal design*. Available at https://www.jal.com/en/ud/haneda.html, accessed 03/03/16.

Japan International Cooperation Agency (JICA) 2009. *JICA Thematic guidelines on disability*. Available at http://www.jica.go.jp/english/our_work/thematic_issues/social/ pdf/guideline_disability.pdf, accessed 03/03/16.

Jensen, C. B. 2007. Infrastructural fractals: revisiting the micro – macro distinction in social theory. *Environment and Planning D: Society and Space*, 25: 832–850.

Jullien, F. 2014. *On the universal: the uniform, the common and dialogue between cultures*. Bristol: Polity Press.

Kawauchi, Y. 2009. *Universal design: a reconsideration of barrier-free*. Boston, MA: Institute for Human Centred Design.

Knowles, C. 2014. *Flip-flop: a journey through globalisation's backroads*. London: Pluto.

Laclau, E. 1996. Universalism, particularism and the question of identity. In *Ethic premises in a world of power*. Ed. E. Wilmsen and P. McAllister, P. Chicago: University of Chicago Press, pp. 45–58.

Lid, I. M. 2013. *Universell utforming. Verdigrunnlag, kunnskap og praksis*. Oslo: Cappelen Damm Akademisk.

London Legacy Development Corporation (LDDC) 2013. Inclusive design standards. London, LLDC. Available at https://queenelizabetholympicpark.co.uk/~/media/lldc/policies/lldcinclusivedesignstandardsmarch2013.pdf, accessed 03/03/16.

McCann, E. and Ward, K. (Eds) 2011. *Mobile urbanism: cities and policymaking in the global age*. Minneapolis: University of Minnesota Press.

McFarlane, C. 2011. *Learning the city: knowledge and translocal assemblage*. Oxford: Wiley Blackwell.

Morita, A. 2013. Traveling engineers, machines, and comparisons: Intersecting imaginations and journeys in the Thai local engineering industry. *East Asian Science, Technology and Society*, 7: 221–241.

Mosko, M. 2005. Introduction: a (re)turn to chaos: chaos theory, the sciences and social anthropological theory. In *On the order of chaos: social anthropology and the science of chaos*. Ed. M. S. Mosko and F. H. Damon. London: Berghahn Books, pp. 1–46.

Mosko, M. S. and Damon, F. H. (Eds) 2005. *On the order of chaos. social anthropology and the sciences of chaos*. London: Berghahn Books.

Peck, J. and Theodore, N. 2010. Mobilizing policy: models, method and mutations. *Geoforum*, 41: 169–174.

Peck, J. and Theodore, N. 2012. Follow the policy: a distended case approach. *Environment and Planning A*, 44: 21–30.

Robertson, R. 1992. *Globalization: social theory and global culture*. London: Sage.

Rowe, D. 2012. The bid, the lead-up, the event and the legacy: global cultural politics and hosting the Olympics. *The British Journal of Sociology*, 63: 285–305.

Sensalis, G. 2014. *Japan in bid to improve accessibility ahead of 2020 Olympic Games* Available at http://www.reducedmobility.eu/20141120534/The-News/japan-in-bid-to-improve-accessibility-ahead-of-2020-olympic-games.html, accessed 03/03/16.

Strathern, M. 2004 [1991]. *Partial connections* (second edition). Walnut Creek, CA: AltaMira Press.

Strathern, M. 2009. Using bodies to communicate. In *Social bodies*. Ed. H. Lambert and M. McDonald. New York and Oxford: Berghahn Books, pp. 148–170.

Temenos, C. and McCann, E. 2013. Geographies of policy mobilities. *Geography Compass*, 7: 344–357.

Tsing, L. A. 2005. *Friction: An ethnography of global connection*. Princeton, NJ: Princeton University Press.

United Nations 2006. *Convention on the rights of persons with disabilities*. Available at http://www.un.org/disabilities/convention/conventionfull.shtml, accessed 3/2/16.

Wagner, R. 1991. The fractal person. In *Big men and great men: The personifications of power in Melanesia*. Ed. M. Godelier and M. Strathern. Cambridge: Cambridge University Press, pp. 159–173.

Wagner, R. 2001. *An anthropology of the subject: holographic worldview in New Guinea and its meaning and significance for the world of anthropology*. Berkeley: University of California Press.

Watt, P. 2013. 'It's not for us'. *City*, 17: 99–118.

11 Artefacts, affordances and the design of mobilities

Ole B. Jensen, Ditte Bendix Lanng and Simon Wind

Introduction

Mobilities research is taking a 'material turn' in which artefacts and sites are included in an emerging understanding of the complexity and multiplicity of everyday life mobilities. Based on selected literature on materialities (Ingold 2011, 2014; Latour 1996, 2005; Yaneva 2009) as well as our own recent work on 'mobilities design' (Jensen 2013a, 2013b, 2013c, 2014; Jensen and Lanng forthcoming; Lanng 2014, 2015; Lanng *et al.* 2012) this chapter adds to the 'mobilities turn' (Cresswell 2006; Urry 2007) some fundamental insights into how design decisions and interventions across multiple disciplines and professions (from traffic engineering over architecture to interaction design) shape the situational affordances of everyday mobility practices. By looking into the 'situational' framework for understanding mobilities (Jensen 2013a) we wish to bring mobilities research and design disciplines and practices closer together. The theoretical background of actor-network theory and non-representational modes of thinking (e.g., Anderson and Harrison 2010; Farías and Bender 2010; Latour 2005; Thrift 2008; Vannini 2015) is connected to a framework of pragmatic situational analysis in combination with the notion of 'affordance' (Gibson 2015; Degen *et al.* 2010; Heft 1988, 2010; Scarantino 2003; Yaneva 2009). Drawing upon this conceptual framework, we utilise the example of a research study focusing on a parking lot in a 1970s suburban district in Aalborg, Denmark (Lanng 2015). We demonstrate the analytical value of adding design to mobilities research; and aim to show how experimental and practical design interventions may help uncover new potential for supposed 'non-places' (Augé 1995). The general objective is therefore to bring together mobilities and design in an attempt to illustrate how many of the mundane sites hosting mobility have the potential to become more than mono-functional and instrumental sites (e.g., parking only). Moreover, this connects to the basic proposition that mobility is more than movement from A to B. By extension we argue that the sites hosting mobilities also have potential for more socially inclusive practices, ludic engagements and affective and sensorial experiences.

Situational mobilities design

As indicated above we take as our point of departure the notion of 'mobilities in situ' (Jensen 2013a). The theoretical underpinning of such an approach is anchored in an interest into the 'real' situations and the multiple small interactions between humans and non-humans that make up the mobile everyday life of billions of people. The specific situation is the place to focus in order to unpack the meaning of mobilities and to understand how various social 'others' (architectures, technologies, infrastructures, service systems, etc.) afford the particular situation. Such an approach builds on the pragmatic question: which architectural artefacts make this mobile situation possible? Our analytical perspective derives from Jensen's (2013a) 'staging mobilities' framework (Figure 11.1), according to which any given mobile situation may be unpacked into three analytically separate spheres: material spaces, social interactions and embodied performances.

For instance, a mobile situation such as riding the subway incorporates elements of materiality related to trains, tracks, platforms, compartment design, gateways, ticket systems, CCTV systems, station architecture, etc. Furthermore, a subway ride is a social event where interaction with fellow passengers, train stewards, newspaper agents, coffee shop attendants, maintenance people and subway

Figure 11.1 The staging mobilities framework (Jensen 2013a, 6).

police, for example, make up a complex social and mobile geography. Last but not least, a mobile practice such as subway riding is an embodied performance in which the capacities and competencies of an individual mobile subject's body determine the ways in which mobile practice is carried out as well as relying upon multiple sensory engagements with subway driving (from sights over smells and sounds to haptic enrolments with train seats or leaning devices). Such a mobile practice (analytically broken down into specific situations) has to be understood on the background of plans, strategies, design manuals, policies etc. all staging the experience 'from above' as it is termed. But equally important is the way we 'carry' our bodies and the multiple incremental decisions we take on the fly in the situations of everyday life mobilities. These are choices and practices made in accordance with the mobile subject's intentions, wishes, affects, moods and preferences. In short these mobile practices are staged 'from below'. By looking at mobile situations as being 'staged' in material settings full of mobile consociates and by mobile bodies we get an operational analytical framework. The framework rests upon an understanding of sites as defined relative to other sites, and further from this relational and mobility-oriented understanding of place also to a relational understanding of the affordances created 'in situ'. Here we move away from a deterministic and 'inside closed-up objects' analysis (Ingold 2011, 16) towards an open-ended enquiry of what actualises the affordances (Shields 2010, 297). In bringing this situational and pragmatic approach to mobilities research we come much closer to issues related to design and architecture. One important reservation should be made here. Even though we are interested in sites and buildings we wish to open up much more broadly to any artefact of relevance in shaping the actual situation. Hence we wish to emphasise a broader notion of 'design' that includes fields such as engineering practices, software and service design. Confining ourselves to architecture 'proper' would miss out on a large number of factors shaping everyday life mobilities. This pragmatic and situational approach partly makes us capable of addressing design issues more widely, but it also opens up dialogue with types of research more oriented towards performances and practices. Such a research agenda shares a number of common traits with non-representational approaches (Anderson and Harrison 2010; Thrift 2008; Vannini 2015). Connecting with non-representational modes of research, the situational approach to mobilities, in general, and its focus on material design, in particular, leads us towards the notion of 'affordance'.

Thinking with affordances

In order to analytically approach the dynamic human–environment relations in mobile situations, we wish to advance the theoretical concept of affordance. With inspiration in pragmatism, we see the theoretical concept of affordance as a tool, an analytical (and design interventionist) device, in which relevance and validity is measured in what it enables us to do, how it aids us in thinking about and elucidating the subject matter in a particular manner. In the context of the design of the physical environment, the notion of affordance, as an analytical and interventionist tool, is interesting in several ways, two of which we unpack below. First, the notion of affordance carries a strong material focus that allows us to take seriously the role of the physical environment and architectural artefacts in mobile situations. Second, it hones our attention in upon the dynamic physical and affective interplay between the material environment and its users. And in doing so, the concept of affordance allows us to become attuned to the performativity of the material world, what it 'offers' and what it 'does'.

In *The ecological approach to visual perception,* Gibson (2015, 119) proposes that "The affordances of the environment are what it offers the animal, what it provides or furnishes, either for good or ill". Through this concept, Gibson (2015) elucidates the possibilities that the environment offers to an organism, for instance, air affords breathing, water affords swimming and a flat surface affords walking. Hence, affordances are "all 'action possibilities' latent in environments, objective measurable and independent of the individual's ability to recognize them, but always in relation to the actor" (Gibson 1977, 67). Consequently, affordances are not mental constructs that we subjectively bestow upon the physical environment; they are also something that the environment offers "because it is what it is" (Gibson 2015, 130). These inherent possibilities of the physical environment should be understood with reference to the organism (i.e. a human being) and its capabilities.

Expanding Gibson's (1977, 2015) conceptualisation of affordances, Heft (2010) argues for an action-based approach. This entails that perceiving what the physical environment offers is not a uniform process in the minds of people, but rather is going on "where the action is", in the specific dynamic coupling of a human body and the physical environment (Heft 2010, 29). Hence, Heft (2010) urges us to remember the idiosyncratic relationships between people, their experiences of a physical environment and its affordances. While Gibson (1977, 2015) focuses on functional aspects, Heft (2010, 22) proposes that the concept of affordances can also be used to tease out "interrelated qualities" of the physical environment of which function is only one. Hence, Heft (2010) proposes to use the affordance concept to highlight 'meanings', thereby addressing the inherent multiplicity of the physical environment due to different (groups of) people's engagements with it. As we shall see in the following section, different people interact with the same physical environment in differing ways, reading widely different meanings and affordances into it.

Furthermore, in addressing this pluralism emerging in the coupling of physical setting and its users, Heft (2010) points to what he calls 'attraction'. With this he emphasises the affective quality or emotional intensity of a possibility in a physical environment. The coupling between users and physical environments or architectural artefacts is never neutral but charged with an affective pull or push (Heft 2010). This is particularly visible in the ways that children engage with their surroundings through playful practices. A bench offers the possibility of sitting, but to many young children it entices them to crawl, climb and jump off it. The bench is not neutrally inviting but almost seductively appealing to certain practices. Likewise, certain mobile situations might draw people in exactly because of the specific affective experiences they may afford, for instance cycling very fast down a hill may hold a specific and emotional "value in engagement" (Heft, 2010, 25). This allows us to move beyond Gibson's notion of affordances as "properties of the environment at which individuals gaze indifferently" (Heft 2010, 25) and to use the concept of affordance to address "meaningful, value-rich features of experience that in the course of action and in the context of an individual's history are often alluring, and sometimes repelling" (Heft 2010, 26). Expanding affordance to include meaning and attraction helps us to think about the environment not only as composed of physical properties but also as meaningful, affective, embodied and idiosyncratic "arenas for action" (Heft, 2010, 28).

In order to theoretically approach such 'arenas for action' and how affordances emerge in the dynamic interplay between environments and users in mobile situations we turn to actor-network theory. Akrich and Latour (1992, 261) describe the notion of affordance as denoting what a device allows or forbids an actor (human or nonhuman). Through the generalised symmetry principle, Latour (2005) urges us to think of the physical environment, artefacts and objects as having agency. This analytical gesture is not meant to evoke the intentionality of things as such, but rather to emphasise their relational capabilities and performative effects because things and humans might affect each other with equal force.

Following this understanding, but in the context of the physical environment, Yaneva argues that architectural artefacts are in no way passive but might very well incite and entice action; and furthermore that design actively "shapes, conditions, facilitates and makes possible everyday life sociality" (2009, 280). Yet, affordances in this perspective are not one-way effects emerging from the physical environment or architectural artefacts structuring human action, but rather they are potentialities effectuated in the assemblages of humans and nonhumans. Shields (2010, 297) captures this symmetry in the formation of affordances:

> For pavement, you can walk on it; you can sit on it; you can drive on it … […] You have to actualize it as this or that. What will it be? It is your choice at any given time. So, in the actualization, people play essential roles. But one should not underestimate the materials: their hardness, their softness, their ability to maintain a shape. All this makes the material a player in a way that is significant, causative, but not causal.

Based upon this we propose to move beyond solely thinking about the physical properties of architectural artefacts and what they might permit or prescribe in an isolated manner and instead focus on the assembling of heterogeneous bodies that, when brought together in the right configuration in mobile situations, might actualise certain affordances for action. Our attempt to wed the notion of affordance to an actor-network-inspired approach allows us to conceptualise affordances as effects of socio-material assemblages. This analytical move offers us a way to carefully dissect and understand the conditions from which affordances are produced and sustained (and in turn provide insight into how they might be designed). Based on these theoretical insights from Gibson (1977, 2015), Heft

(1988, 2010) and actor networks, we are now equipped with a conceptual framework to analytically unfold three mobile situations. These situations demonstrate key aspects of the multiple, relational and affective quality of affordances.

Parking lot affordances

To illustrate the affordances of architectural artefacts we now turn to an ordinary parking lot (Figure 11.2). From this parking lot we draw out several examples of affordances which showcase the multiple and relational quality of architectural artefacts in mobile situations. With the first mobile situation we exemplify the multiplicity of practices and experiences that the same architectural artefact might afford to different users. With the second situation we expand this understanding to also include that the artefact–user coupling is always affectively charged (although with varying intensities) and that this is an essential dimension of the concept of affordance that we wish to advance. The third and final mobile situation draws attention to the importance of analytically re-contextualising the artefact–user relation in its wider physical setting when considering mobilities design. The examples are drawn from a larger research study on mobilities design of ordinary transit spaces in the Danish suburban district of Aalborg East (Lanng 2015).

Aalborg East is a 1970s district, planned and built in accordance with principles of urban functionalism. This entails the zoning of functions, so that housing, industry, recreation and transport are separated from one another in a horizontal and spacious urban plan. The transport function takes up vast areas and is fundamentally organised through segregation and separation of modalities. This extensive

Figure 11.2 The parking lot, November 2011.

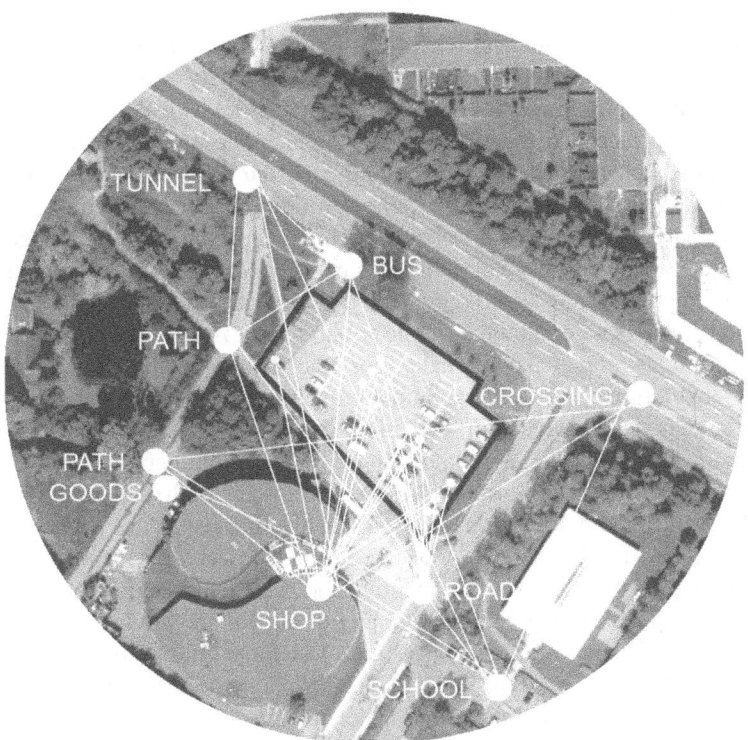

Figure 11.3 Crossing the parking lot.

asphalted system of mobilities was developed to handle the huge increase of cars during the last century, with safety and efficiency as the main concerns. This is the city of the car. And it is the city of surface parking.

At the surface parking lot the segregated mobility system of mono-modality (car drivers and pedestrians/bicyclists respectively) is superseded by a mixed situation. Although the inhabitation presents itself as a very sparse spectacle, the parking lot is, over time, traversed by flows of pedestrians, car drivers, bicyclists and other travellers. They use the open surface to reach their destinations (school, bus stop, shop, parked car, road, pathway, etc.: see Figure 11.3). These travellers form a diverse group of people (in vehicles). During their journeys across the parking lot surface they engage in various mobile situations. For our purpose here of illustrating the affordances of architectural artefacts we begin with this surface.

It is a large planar asphalt surface, levelled out and with white markings for 100 parked cars. It is around 2,600 m², approx. 65 × 40 metres. Which mobile situations does this ordinary artefact afford? Our answer draws upon ethnographic research into two daily life journeys. One journey is performed by Pia, a woman in her mid-30s. The other journey is performed by two schoolgirls, Randa and Diana, aged 13 and 14.

Pia allowed us to travel with her on her route home from work in August 2013. On this journey she gets off the bus at a bus stop immediately above the parking lot. From here she continues by foot across the parking lot asphalt surface. The walk across this surface is a meandering route, which she navigates according to the various obstacles of parked and moving cars, the sparse urban furniture, and other travellers. Pia is a slow walker. She is mildly disabled, and this makes her markedly sharpen her attention to her surroundings on the lot. It is an open, extended surface where cars can come from many directions and move suddenly, she finds. This is a rather unsecure situation that she must navigate with great attention and care. It is an embodied mobile situation that is afforded, in part, by the surface. While the planar asphalt makes possible the obvious functional parking and easy circulation of cars, its continuous expanse also affords that these cars are allowed anywhere and spread out. This in turn produces very little friction that can organise the routes of cars, and almost no places where vulnerable travellers, like Pia, are shielded. This example demonstrates, following both Gibson's (1977, 2015) and Heft's (1988, 2010) arguments, that the affordances of this parking lot surface must be understood in relation to the mobile subject who inhabits it. In the particular mobile situation of Pia's attentive and careful navigation across the lot we see one instance of the dynamic coupling of human and environment that is shaped in her specific practice and experience of traversing this particular parking lot. Through this we get a sense of this banal asphalt surface's performative efficacy, that it can 'do' various stuff, so to speak, in different situations: it affords easy driving and parking for some, while simultaneously difficult and insecure movement for others. This artefact is not just a 'closed-up object' (Ingold 2011) and its affordances are not just 'inside' of it. They are rather relational and multiple: they are brought to life, actualised, in the specific situational relationship with users. And in these relationships the asphalt surface affords various, ambiguous mobile situations. This multiplicity and ambiguity of affordances of this artefact are further demonstrated in the next mobile situation we wish to unfold, which also emphasises affective qualities and emotional intensities. This is what Heft (2010) refers to as attraction that may reside in the dynamic relationship between the environment and its user.

It is not only for car-driving and parking that the asphalt surface's extent and smoothness is constitutive of a convenient affordance. Two schoolgirls, who allowed us to travel with them on their daily route home from school across the parking lot in June 2012, recall how the surface can invite ludic activities. When they were younger they used to come there and enjoy this place, particularly at times when no or only a few cars were there, as the surface gave them the opportunity to play on wheels: "We used to come here and just cycle or skate or something like that, and just go round. It was great, because it's such a large place [...] such a large planar surface, there aren't any pebbles" (Randa, follow-up interview, 5 July 2012). As Randa asserts, the simple characteristic of the smooth, levelled asphalt surface affords not only parking, car-driving, and easy push of shopping trolleys, but also playful activities. What we see in this situation is the specific interplay between two young girls and the surface. The girls read not only a practical possibility, but also this environment potentially holds certain embodied and sensorial experiences for the girls which lure them in. The surface acts as an evocative invitation to engage with it in ludic ways. In the quotation, Randa explains how the strong embodied tactile encounter with the largeness and smoothness is what creates the affective attraction. These are physical properties that come to life in the affective meeting with the girls in a particular mobile situation.

With these examples we advance the understanding of relational, affective and multiple qualities of affordances of an architectural artefact that emerge in the meaningful encounters with its users. Furthermore, this parking lot shows us another point of great importance to design: that the asphalt surface cannot be understood as a discrete architectural artefact isolated from its wider physical context. Its affordances depend on the entire assembling of the setting, herein the relative location to other spaces and artefacts. When the two girls inhabit the parking lot through ludic practices, they do it as part of their daily lives in a network of buildings, urban spaces and routes. This parking lot surface is not just any parking lot surface; it is a specific place in itself but also related to its surroundings, from which the actualisation of its affordances cannot be separated. This point is revealed even more clearly in the last mobile situation that we will bring forth here: a long goodbye on the edge of the parking lot.

Often, when the two schoolgirls walk home from school, they are inclined to get caught in conversation with co-travelling friends. This tends to happen at crossroads, i.e. places where routes intersect and where they have to continue in different directions. One important crossroad is on the very edge of the parking lot. The girls like to linger at this place, delving into a long goodbye, talking together, not really motivated to leave the place. Randa vividly explains the simple presence of the crossroads as affording how the parking lot edge is a place of lingering: "It's right here that we have to go in different directions [...] so it's right here that we gather" (follow-up interview, 5 July 2012). This tiny spot of dynamic proximity on the parking lot edge becomes a site for extended goodbyes, and it is also a place for chance encounters between people who come from different directions. Through this point travellers come and go to and from their parked cars, they are on the way to the bus stop, or to the pathway west of the parking lot, and yet others to the grocery shop. This allows us to see how affordances are co-constituted in the wider physical setting of the architectural artefact. The presence of important local functions (shop, school, etc.) and of transit spaces and routes other than the parking lot itself sets the scene for travellers to actualise the parking lot as an ambiguous public space of meetings, fun, navigation, fast and slow commutes, practical parking and much more.

Through these three empirical examples we wish to elucidate the multiplicity and complexity involved in producing and actualising mundane affordances of an architectural artefact. By deploying actor-network thinking in this analysis, we can analytically juxtapose the users and the asphalt surface with the wider material conditions that co-constitute the mobile situation as an arena for potential action. Not only does this allow us to consider affordances as the performative effects of larger heterogeneous assemblages, but such a conceptualisation also holds value for our larger research ambition of coupling affordances with design thinking.

Conclusion and perspective

In this chapter we have argued for an understanding of both artefacts and their affordances as relationally dependent on the multiple actors enrolled into specific mobile situations. Moreover, we have pointed to the mundane sites of everyday life mobilities as settings where the analytical gaze based on this line of thinking may unfold 'unseen' potentials and meanings. With Heft's (2010) notion of attraction we have become attuned to the affective dimensions of affordances. This resonates with non-representational thinking and Ingold's (2011) request for a refocus from 'the material' towards tangible 'materials' in order to sharpen attention to the multi-sensorial engagements with felt and embodied experiences. Rather than focusing on 'the material' in the abstract, we are centrally concerned with looking at the specific artefact and its relational linkages with multiple relevant 'entities' defining the mobile situation. Even though parking lots tend to be bemoaned as people-hostile asphalt environments, our account of the affordances of the banal artefact of an asphalt surface points to the possible ambiguity and assembled quality of such unheeded places. This attention to affordances suggests a potential for working with the relational capacities of designed artefacts of mobilities to facilitate a wide range of embodied experiences and practices. In the parking lot with which we have been occupied here, this could, for example, point to the potential of addressing, by design, a plurality of mobile situations that occur – or could occur – in the parking lot: to target the affordances of parking lot artefacts, thereby inviting and encouraging social and sensorial everyday life mobilities that take place in variations of modalities, speeds, routes and rhythms.

From the overview of the theory of situational mobilities, the key qualities of affordances and the empirical illustrations of mundane everyday life mobilities, we see a number of interesting perspectives emerging. For one thing, our approach suggests that asking 'which architectural artefacts make this mobile situation possible?' is a pragmatic and straightforward way of addressing affordance as an operational concept that can aid in understanding and working with mobilities design. We see an opportunity for applying the notion of affordance in the mobilities research field as a facilitator pointing towards the emergence of the new field of mobilities design. Many routes may be taking form here, but we would advocate a situational focus on the 'little practices and dramas' of everyday life mobilities. We may also see the fulfilling of a larger goal, namely that of adding to understanding of how we can design meaningful and functional everyday places for people with experiential and affective value. This seems like a standard quest for architects and designers that only recently has become an issue within more academically anchored mobilities research. In our understanding the merging of mobilities and design point not only to new analytical insights and approaches to mobilities research but also to new ways of engaging with practice and with the professions shaping the actual situations for people in mundane settings across the world. The promise of mobilities design may very well be to engage with interventions, proposals for material change and real transitions. These are activist dimensions of knowledge production that once seemed obvious but which for long has been dismissed from academia with its more abstract and theoretical ambitions. Mobilities design may pave the way for new engagements between research and design practice. Finally, we see mobilities design and its conceptualisation to point less towards functionalities, and more towards areas of attention derived from non-representational perspectives. These include an increased attention to the affective dimension of everyday lives such as sensations, atmospheres, embodied and emotional experiences. These are early days for claiming a firm conclusion of where this may lead, but in this chapter we have aimed to illustrate how the merging of a situational understanding of mobilities with a non-representational interest in affordances and artefacts is full of future potential.

References

Akrich, M. and Latour, B. 1992. A summary of a convenient vocabulary for the semiotics of human and nonhuman assemblies. In *Shaping technology/building society: studies in sociotechnical change.* Ed. W. Bijker and J. Law. Cambridge, MA: MIT Press, pp. 259–264.

Anderson, B. and P. Harrison (Eds) 2010. *Taking-place: non-representational theories and geography.* Farnham: Ashgate.

Augé, M. 1995. *Non-places: an introduction to supermodernity.* London: Verso.

Cresswell, T. 2006. *On the move: mobility in the modern western world.* London: Routledge.

Degen, M., G. Rose and Basdas, B. 2010. Bodies and everyday practices in designed urban environments. *Science Studies*, 2: 60–76.

Farias, I. and Bender, T. (Eds) 2010. *Urban assemblages. How actor-network theory changes urban studies.* London: Routledge.

Gibson, J. J. 1977. The theory of affordances. In *Perceiving, acting, and knowing.* Ed. R. E. Shaw and J. Bransford. Hillsdale, NJ, Lawrence Erlbaum Associates, pp. 127–143.

Gibson, J. J. 2015. *The ecological approach to visual perception.* New York: Psychology Press. (Originally published 1986.)

Heft, H. 1999 [1988]. Affordances of children's environments: A functional approach to environmental description. In *Directions in person-environment research and practice.* Ed. J. Nasar and W. Preiser. Aldershot: Ashgate, pp. 43–69.

Heft, H. 2010. Affordances and the perception of landscape. In *Innovative approaches to researching landscape and health.* Ed. E. C. W. Thompson, P. Aspinall and S. Bell. London: Routledge, pp. 9–32.

Ingold, T. 2011. *Being alive: essays on movement, knowledge and description.* New York: Routledge.

Ingold, T. 2014. Designing environments for life. In *Anthropology and nature.* Ed. K. Hastrup. London: Routledge, pp. 233–246.

Jensen, O. B. 2013a. *Staging mobilities.* London: Routledge.

Jensen, O. B. 2013b. Mobility divides – 'staging' differential mobilities. Keynote paper for the 4th PanAmerican Mobilities Network Conference 'Differential Mobilities', Concordia University, Montreal, 8–13 May.

Jensen, O. B. 2013c. Designing mobilities – staging materialities of mobilities. Paper for the Danish Sociology Congress 'Mobilitet & By', Roskilde University, Denmark, 24–15 January.

Jensen, O. B. 2014. *Designing mobilities*. Aalborg: Aalborg University Press.

Jensen, O. B. and Lanng, D. B. forthcoming. *Mobilities design: urban designs for mobile situations*. London: Routledge.

Lanng, D. B. 2014. How does it feel to travel through a tunnel? Designing a mundane transit space in Denmark. In: *Ambiances [Online], Experimentation – Design – Participation*. Online since 15 October 2014, connection on 23 October 2014. URL: http://ambiances. revues.org/454

Lanng, D. B. 2015. *Gesturing entangled journeys: mobilities design in Aalborg East, Denmark*. PhD dissertation. Aalborg University, Denmark.

Lanng, D. B, Harder, H. and Jensen, O. B. 2012. Towards urban mobility designs: en route in the functional city. Paper for the conference Trafikdage, Aalborg, 27–28 August 2012.

Latour, B. 1996. *Aramis or the love of technology*. Cambridge MA: Harvard University Press.

Latour, B. 2005. *Reassembling the social*. Oxford: Oxford University Press.

Scarantino, A. 2003. Affordances explained. *Philosophy of Science*, 70: 949–996.

Shields, R. 2010. Interview with Rob Shields. In *Urban assemblages. How actor-network theory changes urban studies*. Ed. I. Farías and T. Bender. London: Routledge, pp. 291–301.

Thrift, N. 2008. *Non-representational theory: space, politics, affect*. London: Routledge.

Urry, J. 2007. *Mobilities*. Cambridge: Polity Press.

Vannini, P. (Ed.) 2015. *Non-representational methodologies: re-envisioning research*. London: Routledge.

Yaneva, A. 2009. Making the social hold: towards an actor-network theory of design. *Design and Culture*, 1: 273–288.

Part III

Design knowledges

Making connections

12 Towards a new discipline

The design of urban vehicles

Lino Vital García-Verdugo

[T]oda cosa grande para la vida, es ensoñada primero por un poeta; traducida a honda prosa al instante por un filósofo; hecha tangible luego por un científico analítico; democratizada y puesta al alcance de todos por un industrial, y por el industrial, al fin, empequeñecida.

(Roso de Luna 2006, 26–7)

Every great thing in life is dreamt by a poet first; instantly translated into deep prose by a philosopher; then, materialised by an analytical scientist; democratised and made accessible to all by an industrialist; and, by the industrialist too, ultimately belittled.

(Author's translation)

Introduction

In 2015 the motoring press was extolling the virtues of the latest sports car designed by a mainstream manufacturer (Quiroga 2015): a low-powered machine able to 'thrill' with its clever design (lightweight and responsive chassis). This 'revolution' was actually an update of a 1989 model, which itself took inspiration from English design engineer Colin Chapman's 1962 Lotus Elan. In effect, one of the most coveted cars of 2015 was an update of a 53-year-old concept.

What might otherwise be a footnote in automotive history provokes considerations about the role of ageing automotive concepts in congested twenty-first century cities. It is equally interesting to see how Chapman's ideas—"Simplify, then add lightness" (Lotus 2016), and focus (cars as design motif) are still relevant. When one can commute in a sedan nearly as powerful as a Formula One car, his 'frugal' design ethos is scarce yet timely. Chapman belonged to an era of automotive evolution where individuals developed visions of private mobility. The search for affordability, driving experience or reliability increasingly focused on the car as an end product, even when it originated as a tool for mobility. In the current era, when the aggregation of individual convenience provided by these highly evolved machines results in gridlock, pollution and daily fatalities, design is confronted with—and indeed *should* be confronted with—a different reality.

When new technologies allow unthinkable opportunities to redefine our environment, it might be time to re-evaluate our quest for mobility. Cars have an undeniable appeal: the capacity to convey emotions by moving in easily accessed infrastructures (roads) is difficult to equal. As such, it seems plausible to see a future of automotive toys. However, beyond mobility-related hobbies, there is another design dimension: cars as physical entities which condition daily environments. So commonplace in most cities that they become taken for granted, these metal boxes not only alter spaces with their movement by introducing hazards and pollution, but also cars occupy and mould interaction with wider spaces.

From a designer's perspective, there is an enormous opportunity in switching the focus from the car as artefact to the private and public interactions it can enable. The multidimensionality of the issue (cars simultaneously can create public disturbance and emphasise private choice, and are mass-produced products) requires interdisciplinary approaches to apprehend the situation and contribute at the intersection of different worlds. This chapter reflects on the role of design and its influence in the re-humanisation of urban environments through unexpected sources: traditionally individualistic vehicles. It starts with a brief overview of how the car evolved from a human-centred solution to the money-maker of today, and continues with an overt criticism of the design practices derived from this industrial reality. From this 'debris', the text makes a brief incursion into a new design approach and seeks to illuminate the ambivalent nature of vehicles (private–public) both as a challenge and opportunity for new creations. Finally, the chapter points towards a new scheme, where mobility affordances will be integrated with contextual values in a way which might create re-humanised and dynamic environments. Let us start by depicting the evolution of cars in their role as unintended moulders of our environments.

Unexpected urban elements

In addition to creating pollution, congestion and danger, cars destroy urban spaces by design. Their imposing volumes divide and conquer local areas, with no other control than equally conspicuous road markings and basic driving skills. Moreover, the combination of unrelated and competing designs (aimed at showrooms) produces chaotic streetscapes. In cities where the minutiae of design from fences to facades are closely planned and regulated, cars enjoy unequalled levels of design freedom. Even the upcoming implementation of technologies aimed at reducing cars' environmental and social impacts (i.e. zero tailpipe pollution, autonomous driving) seems to emphasise the commercial product over urban habitats. From a design perspective, the question of how cars have evolved from a solution to human needs to become urban problems is a useful point to start considering desirable alternatives.

A brief depiction of car design evolution

Cars are *the product of organisations with the sole goal of profiting by selling them*, at least, the cars relevant for this analysis: mass-produced machines. There are passionate masterpieces (cherished in garages and museums) but the ones contributing to urban landscapes are commercial products. This idea has significant implications for two areas related to their ideation: *authorship* and *design realm*.

By the turn of the twentieth century, the car was a novel device addressing and reshaping human mobility needs in relation to comfort, convenience, affordability and distance. Engineers were still figuring out how to fit engine, luggage and occupants onto a self-propelled platform (a very exciting time for automotive tinkerers). Once a basic layout was established as a valid solution, a fundamental change came: the machine should not only transport people and goods, but had to be *designed for mass-scale manufacturing* too. Henry Ford was probably the most successful in figuring out this 'hidden' dimension of the car, to the point of choosing colour range based on paint-drying performance (Votolato 2015). Considering the complexity of tasks and resources involved, that change meant the beginning of the car as the result of an organisation of specialists. Machines created by individuals continued, but it was the *designed-by-committee vehicle* with which most people would live in urban areas.

In parallel, automotive technology continued to evolve through both collective and individual contributions. Aerodynamics and metal stamping triggered the second leap in the materialisation of cars as urban elements (visual and spatial). The car lost its horse-drawn influences to become the product everybody got to know: a *prismatic volume of metal and glass* with a specific arrangement of openings, windows and assorted detailing. Technical evolution would continue fostered by markets and regulations, but the overall typology of the car—and its design realm—dates back to the late 1940s, when the body became a flush surface (General Motors 1958). From that moment, the vehicle became a reflection of itself, incidentally providing mobility, fundamentally fostered by commercial profitability. A good example of the self-replicating nature of vehicle design was set in stone in the late 1990s. When BMW was searching for a replacement to the 1959 Austin Mini, two studios competed with their own visions: one proposed a compact car that kept the original spirit in the development of a new small architecture; another proposed a 1990s update of the original style on a larger architecture (Ready 2015). The latter succeeded, and has proved to be a commercial hit. Today's MINIs (even the official name grew bigger) are an oversized caricature of the 1950s design, largely having shed the meanings associated with the original.

Synthetically, the evolution of cars as manufactured objects can be seen to have begun as a response to human needs (convenient and comfortable mobility opposed to other means such as motorcycles). From the moment manufacturers figured out profitable ways of building them, the initial needs were translated to corporate language. The needs transformed into key performance indicators (KPIs), a systematisation representing the criteria potential buyers would use and compare in order to decide whether or not to buy a particular model (Pichler 2015). KPIs permeated technical specifications which allowed greater levels of 'objectivity' in the creation of new cars. A key outcome of such specifications was that they enabled manufacturers to compare one product with a competitor's, establishing leaders and followers. Ultimately, markets adapted to automotive standards, perpetuating the nature of products rather than searching for solutions to human needs. A famous quote, attributed to Henry Ford, could not have illustrated this point better: "If I had asked people what they wanted, they would have said faster horses".

In this Sisyphean picture, design became a tool to keep the status quo, providing a dimension of variety (shallow aesthetics) over common platforms (Sako and Warburton 1999). Similarly to other forms of creation (Adorno 2001), the commercial role of design as employed by automotive manufacturers became constrained by a set of pre-defined KPIs, severing any possibility of responding to human needs. The result has been cities occupied by imposing symbols of material status. How design can revitalise itself to move away from this situation and respond to current human needs in urban environments is the next question this text addresses.

Changing approaches

"You are the Michelangelos [sic] *of our times"*. Among tables filled with sketches, screens and electronic drawing pads, a senior designer lectured Masters students in one of the best vehicle design schools worldwide. Rather than the particulars of such a motivating assertion, it illustrates three relevant points in current vehicle design practice:

- First, vehicle designers have far more influence in our times than Michelangelo ever thought he could have (there is no need to go to museums or Florence to experience vehicle designers' creations first-hand).
- Second, there is a sparkling contrast in profiles: Michelangelo was one of the totems of an age that apprehended past universal knowledge and projected to the future. His creations profited not only from his expertise with the chisel, brush or drawing board, but also from direct interdisciplinary experience. Perhaps more important was his connection to the past of humanity on his way to the future.
- Third, it illustrates the egocentric element present in the discipline, which combined with lack of vision, sensitivity and humanist culture can negatively disrupt the life quality of individuals (the point at hand for this discussion, over any professional criticism).

The approach of vehicle design research has been to combine an apprehension of its ubiquitous influence with a deep understanding of the problems at hand and the tools currently available to produce meaningful solutions. As stated in the introduction, the designer is not the sole author of the car as an urban problem (the manufacturer as organisation ultimately conditions creative inputs). However, the designer is certainly in a position to see the vehicle as an interdisciplinary product conditioning citizens' lives. The contribution of the design professional should be an indirect orchestration of vehicle materiality, rather than one more isolated input to an inert volume.

In the mobilisation of design, the argument in this chapter is one that seeks to disrupt and widen common practices, moving away from the computer screen or clay workshop to the apprehension of the problem at hand. Thus, the approach consists of two separate stages: an analysis of the definition of private and public space in terms of urban movement; and an assessment of current know-how to project this analysis towards future solutions.

The nature of the text responds to two factors: practice-based character of design and existing theoretical analyses. Being a highly applied discipline, vehicle design always relates to solutions, especially when considering disruptive proposals. Analytical theories supporting the nature of alternative proposals already exist; the missing link is the materialisation (even at conceptual levels) between related alternative theory and vehicle design practice.

Mobile privacy, public space

In the design of spaces or artefacts, there is a fundamental (often forgotten) human factor. Rather than partisan approaches perpetuating certain design practices, a focus on the wider picture will highlight how apparent automotive issues are nothing more than reflections of higher-level concerns for individuals and groups. Moreover, such holistic views can help to identify hidden factors fostering disruptive solutions. The contrast of private interests within shared realms will kick-start the definition of new human-centred alternatives around connectivity.

To start studying the private–public conflict in mobility, it is first worth considering the *raison d'être* of cities in their origins: settlements built by individuals to obtain easier access to services. Defendants of public mobility point to improvements in the goods and services provided because centralised solutions are quantitatively more efficient. However, there is a qualitative difference between individuals and livestock in private mobility. Considering mobility as mere resource allocation has clear links with modernist theories that tried to convert houses into machines for living (approaches that directly contributed to contemporary negative side-effects of mobility today). The problem of cities is (now or in an urban future) how to provide services to individuals, rather than headless masses. Clearly, the problem of transportation is key in a future where the world population is expected to concentrate in cities (UN HABITAT, 2013); but logistic considerations should not annihilate the individuality of millions of citizens. The fact that design does not consider individual and social needs is not an obstacle for people to find unexpected alternatives (Staeheli and Mitchell 2006), yet it remains a responsibility of designers (and related policy-makers) to take into account multidimensionality.

Analyses of future mobility starts with a question on the need to move at all. Futurists for example are revisiting Paris and its distributed services as an organic way to provide both convenience and life quality (Larson 2014). Yet whilst computers might help us to re-humanise urban planning, in the meantime, short-term futures still require urban mobility on a daily basis. Even for future pedestrianised cities, this study could serve as an extreme scenario to explore the role of mobility

in the definition of local spaces. Therefore, the questions posed by current challenges would be: *why private mobility?*; and *how does it affect others?*

The private allure of cars

Sociology provides the core idea explaining the preference for cars: a desire for
a *seamless journey* (Urry 2000). Within the chaotic spaces of modern cities, cars
offer controllable transitions from departure to destination (i.e. home to work) both
in terms of space and scheduling. In masterminded and variegated cities, conventional journeys often cross through daunting areas (the archetypical *dark alley*) that
can be avoided with these machines. Moreover, they also preserve personal space,
a key factor of individuality in environmental psychology (Hall 1969). In terms of
time, private transport allows the individual to plan personal schedules, something
especially relevant when several members need to coordinate their trips.

In fact, interactions between family members further reinforce the preference
for cars. While independent citizens can tolerate public transport inconveniences,
those in charge of children or elders may be less able to compromise (Miller 2001;
Boyer and Spinney 2016). Cars become spaces drivers offer to others in daily
routines, often taking the role previously played by homes in family time. In that
sense, the fundamental character of home as a reference space for articulating
caring relationships (Mitscherlich 1965) can also be attributed to cars. As such,
mere vehicles become key elements in the formation of individuals and families
(particularly for children, cars become a vantage point to discover the world from
a controllable environment).

However, the influence of cars as lived spaces does not mean current vehicle
design accounts for the full range of social practices or concerns outside of these:
the main goal of vehicle design is to produce a favourable purchase decision:
it is incidental that cars play fundamental roles in their users' lives. As other
sociological studies have shown (Staeheli and Mitchell 2006), it is not design
intention, but the absence of suitable solutions that provokes a search to satisfy
social needs through commercial creations. In fact, marketing trickery often confronts human needs: intrusive family car interiors artificially separate occupants
within the same shared space! They also foster a separation from travelled space,
for example encouraging children to look at TV screens, oblivious to the view out
of the window. Far from annihilating private mobility, there is a clear opportunity
for design disruptions to significantly improve lives in mobile urban contexts.

Influence of private mobility on public spaces

Cities, considered as habitats, are embodiments of social life over physical spaces.
They introduce immanent qualities with respect to human needs, both with evident (i.e. shelter) and subtle advantages, such as the concept of *genius loci*, the
commonly perceived role of certain sites (Rossi 1982). Until the twentieth century, the origin and growth of urban nuclei mostly responded to typical human
needs. In longer-established settlements, the democratisation of cars as mobility

solutions altered both space use and urban sprawl. In terms of space use, cars not only introduced their spatial needs (roads and car parks), but also adapted local spaces through stance and position. In terms of stance, their nature as status symbols negates any consideration for proportions, harmony or human scale that drives the shaping of urban space in other design disciplines (Hill 2003; Stevens 2007). In addition, eye-catching surfaces and details compete for attention in spaces that were never intended to be car showrooms. Ultimately, these alien elements are irregularly grouped with no other order than parking lanes combined with their drivers' abilities, introducing another paradox where such intrusive volumes escape the control of urban planning authorities: They can create physical and visual walls or just condemn surrounding spaces. It is, frankly, incredible how the 'advantage' of mobility has surpassed the importance of settlements themselves.

Nevertheless, the effects of car use are not an isolated aggression against urban habitats, but an example of misled growth. An uncontrolled technical prowess and land used as an economic asset generate inhabitable spaces. In fact, it has been the same prioritisation of individual interests over the social character of cities which is at the root of urban issues (Mitscherlich, 1965). Perhaps the ubiquitous character of vehicles, and the separation of their use from land ownership places them in a good position as tools of change to reconquer urban spaces.

The dynamism of the car can also be a way to resolve the interaction of traditional spaces to the 'incursion' of a new space: multimedia realities. Multimedia has invaded urban spaces, enabling mobility without physical displacement (Virilio 2004). In addition to the obvious advantages of immediate access to information, new technologies can work against cities too. When there is no need to wander, to walk the city, technologies which supposedly connect can establish barriers and end up contributing to the loss of that sense of belonging that car excesses provoke too (Bratton, 2009).

When cities are becoming a succession of ever-changing screens, both hanging on façades and moving along sidewalks, architects have reflected upon the need to reinstate stable contexts (McCullough 2015), to generate sense-making in a quest to integrate progress in cities to which individuals can relate. Interestingly, the need for references cited in the formation of the individual (Mitscherlich 1965) reappear in the need to re-humanise spaces. One cannot but wonder whether some of the intrinsic qualities of cars (volumes and spaces, static and dynamic) can be used to contribute to a re-imagining of urban spaces.

Materialising mobile intersections of private and public

In the consideration of a new type of vehicle from a human-centred approach, the debate should not revolve around existing product qualities and typologies but around materialising affordances with current means. As the second section illustrated, cars have become 'frozen' design entities; initially providing affordances to users they have reached a moment where the incorporation of new affordances is blocked by their own industrial and marketing inertia. Design in this sense has become a discipline perpetuating past paradigms. Accordingly, the mobilisation of design should focus on the reinvigoration of the process: the incorporation of latent affordances to create something new (not a new car, but a new solution). In that sense, the accepted product qualities of the vehicle should be re-imagined in relation to the provision of new affordances. In this section I will briefly outline what these new affordances might be.

Fundamentally, the main change is the *apprehension of context* in vehicle design. Future proposals would have to be developed around two core ideas: *transport-related affordances* (in common with cars); and *interactivity*. These ideas represent two levels of affordances. The first matches current car design concepts. The vehicle should provide convenient transportation. The second point is a radical departure in that the re-contextualised vehicle can interact with people, other vehicles and spaces.

Transport-related affordances

The architect Frank Gehry's creative approach is to start designing *a* solution that works. An initial design will end up in the bin with the sole purpose of familiaris-ing with the design constraints. Once binned, Gehry's studio focuses on creating 'their' solution (Boland and Collopy, 2004). In times when normal drivers can buy a 700-horsepower sedan car as a daily commuter vehicle (Okulski 2015), con-sidering affordances related to 'mere' mobility can be considered an initial step in the design process of future urban transportation. It does not need to be intrusive with modern technology. The point at hand is how to create a *mobile space* able to manage interactions, and how to use current technology to eliminate distractions in three different levels:

1   *Motion*: State-of-the-art automotive technology can create flat-bed platforms able to move with a subtlety close to a magic carpet (Vital 2012): a sandwich chassis can enclose the main components; electric powertrains can be silent, hidden inside the flat-bed chassis and provide smooth immediate movement without the need of any transmission gear and clutches; self-driving capabili-ties can further contribute to the magic of effortless displacement.
2   *Passive safety*: Cities present varying mixes of traffic, from pedestrian to high-speed roadways. However, trips in cities are typically slow and short (Transport for London 2010). Alongside the introduction of low-speed zones, such an understanding can help to drastically simplify vehicles that are cur-rently over-designed for ways they are used. Alternative typologies such as European quadricycles (European Commission 2010) can be useful both in terms of convenience and the design possibilities (structures can become design resources).
3   *Isolation*: Again, low-speed vehicles and the use of new materials and reshaped surfaces can enable new design possibilities to facilitate interactions.

Automotive technology, in this new view, evolves from being an icon to becom-ing a design commodity directed towards dynamic urban elements.

Interactivity

In the search for a balance between private and public, urban vehicles have to forget the showroom as the locus of design. Despite the timely and still emerging topic of interactions as an object of creation (Kwastek 2015), the design of vehicles is completely oblivious to the possibilities of their imposing nature. Rather, design should embrace their highly interactive and connective position within cities.

From this perspective, the design of urban vehicles can operate in two dimensions: users' location and attention level. Interactions can happen inside, outside or from one space to the other; interactions can also have the vehicle as a motive (foreground) or context (background). Thus, there are five basic types of interactions:

1   When interactions happen between occupants and the vehicle, it becomes a device privately used (i.e. using its infotainment to surf the web).
2   If occupants interact with each other inside, the vehicle acts as a private space.
3   When bystanders interact with the vehicle, it becomes a connective public instrument (i.e. an information point).
4   If bystanders interact with the vehicle alongside other elements of the city, the vehicle becomes a connector to the streetscape and provides a context for other interactions.
5   Ultimately, the vehicle could work as an interface or screen between inner and outer spaces. In that sense, it can vary in its role between foreground and background (i.e. for a conventional car, it would be equivalent to the difference between racing and cruising spaces).

Figure 12.1 represents a simplification to apprehend the different dimensions involved in the design. Clearly, these interactions can be simultaneous, they can change their nature (interaction or isolation), there can be unexpected actors and they are highly dependent on the vehicle speed. Nevertheless, there is a need to design a space that both enables and manages interactions and connections. A concept emerges: the vehicle surfaces as membranes (Vital 2012) fostering an active modulation and interaction of 'in' and 'out'. Over imposing inert volumes, the vehicle as a space able to move is a catalyst for interactions. Its presence, mostly volume, has to be seen as a moulding element with ambivalent nature: a latent volume waiting to be used to enable, to divide.

From the minimal mobile unit described in the previous point, the question would be how to materialise the car as a spatial element able to manage these interactions. Current solutions focus on one actor (the purchaser) and tend to foreground both from outside and inside. Points, lines and planes are arranged in self-reflecting entities well-known as cars, not as interactive elements. In contrast, the formality of a new alternative should be reduced to a level of acceptance (in the end, it will be chosen as private transport).

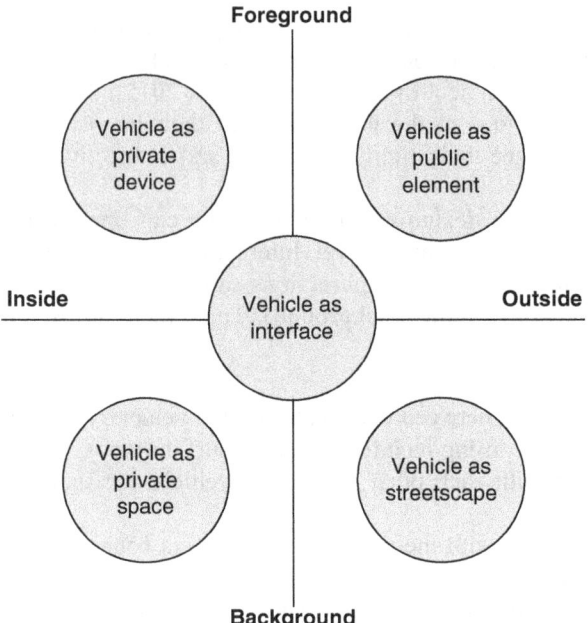

Figure 12.1 The vehicle as a contextualised space for interaction.

In terms of volumes, a basic embracing human-scaled volume would offer a starting point where attention is only focused on the volume as an enclosure of latent features (Vital 2012). For surfaces, alternative disciplines such as architecture are starting to show clever use of transparent materials to produce 'actuated designs', with 'on' and 'off' states that overcome the inert qualities of typical opaque panels (Bruno, 2014). Such lessons would be applicable in mobile machines too. Lights can be used to modulate interaction and isolation. Combined with new materials and multimedia, the design possibilities over apparently simplistic volumes can be endless. They can also be combined with the vehicle architecture to increase appeal with the arrangement of elements in layers underneath the generic container, creating *layered designs* (Vital 2012).

Is this a rupture with past designs? Certainly this approach tries to break with current practices. However, car design understood as material culture is a useful resource. In the creation of private solutions, people accept and are used to interacting with cars as design types. Their lack of novelty minimises the distraction of a subtle car-like volume too. Despite this, however, the utility of an existing visual culture does not place current vehicle design in a better position to conceptualise sensible urban proposals. It is certainly an enticing field for a new kind of design professional, one able to combine the affordances of a mobile space with those of physical and multimedia contexts.

Towards a new discipline

Similarly to the crucial importance of water supply for Roman cities, the city of the future will be enabled by its mobility systems; and private mobility arises as a key element. When pollution, gridlock or road safety emblematise the negative side of cars, it is important to reassess its role in social dynamics. Cars are not only the motor of an industry that employs hundreds of thousands of individuals, but also future cars can redefine the link between individuals and cities. In contrast to smartphones for example, they will connect occupants both to physical and augmented realities of future mega-cities: an instrument to maintain a controllable bubble in otherwise increasingly intrusive environments. Furthermore, future private vehicles can define shared space, opening unthinkable design possibilities such as helping to harness street dynamics as tools to re-humanise cities.

Towards that vision, this study highlights the need for a new design discipline, and a human-centred formal realm for urban vehicles. In terms of discipline, vehicle design has become a mere tool of an industry, a reflection of a past that is not able to conceive anything beyond 'look-at-me' mobile boxes. The problematic of cities and mobility requires new professionals able to combine a humanist vision with state-of-the art tools to propose meaningful solutions (cars or not). Perhaps the incursion of non-automotive industries can motivate a new type of professional with a greater level of freedom to focus on concerns (even if commercially motivated) rather than specific materialisations.

How would this vision combine with mass-made products? As urban furniture, vehicles should be affected by design policies inasmuch as they contribute to urban spaces. What could be seen as a limitation can be the trigger for new possibilities. Urban vehicle design policies should assure the collective convenience of public spaces, while fostering the improvements of those spaces with the possibilities enabled by new types of vehicles. It is certainly a fantastic opportunity to revitalise old disciplines by rediscovering the humanist side.

References

Adorno, T. 2001. *The culture industry: selected essays on mass culture.* London: Routledge.

Boland, R.J. and Collopy, F. 2004. *Managing as designing.* Stanford, CA: Stanford University Press.

Boyer, K. and Spinney, J. 2016. Motherhood, mobility and materiality: Material entanglements, journeymaking and the process of 'becoming mother, *Environment & Planning D: Society & Space* (online before print: doi: 10.1177/0263775815622209).

Bratton, B.H. 2009. iPhone city. *Architectural Design* 79: 90–97.

Bruno, G. 2014. *Surface: matters of aesthetics, materiality, and media.* Chicago: University of Chicago Press.

European Commission 2010. *Proposal for a European Parliament and Council Regulation. Regulation (EU) No .../2010 of the European Parliament and of the Council on the approval and market surveillance of two- or three-wheel vehicles and quadricycles.* Brussels: European Commission.

General Motors 1958. *Styling: the look of things.* Detroit: Public Relations Staff (General Motors).

Hall, E. 1969. *The hidden dimension.* New York: Anchor Books.

Hill, J. 2003. *Actions of architecture: architects and creative users.* London: Routledge.

Kwastek, K. 2015. *Aesthetics of interaction in digital art.* Cambridge, MA: MIT Press.

Larson, T. 2014. City science with Kent Larson. Available at: https://www.architects.org/news/city-science-kent-larson, accessed 30/12/2015.

Lotus Cars 2016. Lotus philosophy. Available at http://www.lotuscars.com/about-us/lotus-philosophy, accessed 10/05/2016.

McCullough, M. 2015. *Ambient commons: attention in the age of embodied information.* Cambridge, MA: MIT Press.

Miller, D. 2001. *Car cultures.* Oxford: Berg.

Mitscherlich, A. 1965. *The inhospitality of our cities.* Frankfurt: Suhrkamp .

Okulski, T. 2015. Ten things you learn after four days in the Dodge Charger Hellcat. Available at http://www.roadandtrack.com/new-cars/first-drives/a26249/dodge-charger-hellcat-test/, accessed 30/12/2015.

Pichler, R. 2015. 10 tips on how to choose the right product key performance indicators (KPIs). http://www.romanpichler.com/blog/10-tips-how-to-choose-the-right-product-key-performance-indicators-kpis/, accessed 04/05/2016.

Quiroga, T. 2015. 2016 Mazda MX-5 Miata: Everything we love about the Miata, but smaller. http://www.caranddriver.com/reviews/2016-mazda-mx-5-miata-first-drive-review, accessed 30/12/2015.

Ready, O. 2015. Concept car of the week: Rover Mini Spiritual and Spiritual Too (1997). http://www.cardesignnews.com/articles/concept-car-of-the-week/2015/06/concept-car-of-the-week-rover-mini-spiritual-and-spiritual-too-1997/, accessed 30/12/2015.

Roso de Luna, M. 2006. *El tesoro de los lagos de somiedo.* Edición de Esteban Cortijo Parralejo. Sevilla: Biblioteca de Rescate, Renacimentio.

Rossi, A. 1982. *The architecture of the city.* Cambridge, MA: MIT Press.

Sako, M. and Warburton, M. 1999. *MIT international vehicle programme. Modularization and outsourcing project. Preliminary report of European research team.* Boston: MIT Press.

Staeheli, L.A. and Mitchell, D. 2006. USA's destiny? Regulating space and creating community in American shopping malls. *Urban Studies* 43: 977–992.

Stevens, Q. 2007. *The ludic city: exploring the potential of public spaces.* London: Routledge.

Transport for London 2010. *Travel in London*, Report 2. London: TfL.

UN HABITAT 2013. *Planning and design for sustainable urban mobility.* London: Routledge.

Urry, J. 2000. *Sociology beyond societies: mobilities for the twenty-first century.* London: Routledge.

Virilio, P. 2004. The third interval. Available at: http://www.paulos.net/teaching/2009/AE/readings/protected/CyberCitiesReader-Virilio.pdf, accessed 30/12/2015.

Vital, L. 2012. µcar: multidisciplinary development of an electric vehicle typology for the city. Available at: http://researchonline.rca.ac.uk/1354/1/VITAL%20Lino%20Thesis.pdf, accessed 05/06/2015.

Votolato, G. 2015. *Cars.* London: Reaktion.

13 Being wheeled through the hospital

Designing for hospital patients' spatial experience in motion

Margo Annemans, Chantal Van Audenhove,
Hilde Vermolen and Ann Heylighen

Introduction

How we move or are moved shapes our experience of the built environment. Thus architects and others designing spaces through which many people move are confronted with the challenge of taking into account people's spatial experience in motion. Locations where patients spend a considerable amount of time moving or being moved, such as hospitals, could benefit significantly from a better understanding of people's spatial experiences in motion. Based on an improved understanding of wheeled patients' spatial experience we look for ways to inform architects' design processes. By providing architects with relatively 'raw' visual and narrative data which have had only limited post-processing (Annemans *et al.* 2012), we aim to gain insight into the impact of different information formats that introduce motion in design. We argue that close consideration of information formats should inspire and trigger architects to focus on spatial experiences in motion, as such creating buildings based on an improved understanding of the connections between mobile subject and space.

The chapter reflects upon a workshop in which participants were asked to design a lift using research data in various formats on hospital patients' spatial experience in motion (Annemans 2015). We analysed the workshop's design outcome and design process, paying specific attention to the affordances of different information formats. The resulting design ideas offer important insights into the relationship between space and motion. The workshop findings illuminate the ways in which design outcomes emerge from the connections enabled by representations of different information formats.

Introducing patients' experience in motion in the design process

Designing for patients' well-being requires an in-depth understanding of their experience. Empathy with users in the design process can be developed in various ways (Kouprie and Visser, 2009). Ideally, designers obtain information through interaction with real target users, i.e. patients, allowing the development of a more thorough understanding, connection and empathy with them (Kouprie and Visser, 2009; McGinley and Dong, 2011). However, since time and money restrictions in a typical design process result in minimal user engagement (Cassim, 2010), designers are often unable to obtain this direct input from users and become dependent upon indirect sources of human information (McGinley and Dong, 2011).

Various techniques have been developed to bring designers closer to users' experience (Kouprie and Visser, 2009; McGinley and Dong, 2011; van Rijn *et al.*, 2011). Most methods aim to foster designers' empathy with those for whom they are designing. The specific situation of particular users also affects the degree to which actual interaction can be achieved. In the case of patients being moved through a hospital, practical and ethical restrictions make it hard for designers to actually engage with them in the situation under study. Therefore, we set out to explore which formats could be suitable to inform design about hospital patients' spatial experience in motion. We aimed to find a format that meets the designers' requirements and was able to communicate data about motion.

The development of a story, like that of a patient's hospital experience, and the trajectory along which it develops, often happen in parallel (Ingold, 2011). An information format aiming at informing design should ideally reflect this parallelism. In a traditional design briefing the parallel development of a patient's story and trajectory is often unclear: there is a tendency to focus on functional and organisational matters. Including user information is often limited to attention to the values of the (care) organisation (Bogers *et al.*, 2008; Elf and Malmqvist, 2009; Elf *et al.*, 2012), without attending to patients' stories. Research suggests that an alternative approach to the design brief with a focus on personal stories (Van der Linden *et al.*, 2016) improves a designer's ability to relate to users' experience. Introducing real users' experiences in the design process allows designers to relate to people's specific situations (Annemans *et al.*, 2014; van Rijn *et al.*, 2011). In product design, co-creation and other forms of designer–user interaction are fairly common (Howard and Somerville, 2014). However, bringing this human-centred approach to architecture and planning is apparently "a big nut to crack" (Sanders, 2009).

Designers are particularly motivated by visual communication and like information to be graphically presented (Lofthouse, 2006). Moreover, they often mistrust data that have already been through a process of interpretation (Restrepo, 2004) and seem to prefer raw data in a format that is condensed to be design-relevant (McGinley and Dong, 2011). These insights, together with the difficulty of grasping the experience of motion in words, mean that visual communication tools would seem to be promising in transferring patients' impressions of moving to architects. While static images can trigger reflection upon motion (Annemans *et al.*, 2012), video seems even better given its mobile character. Video material can be introduced into the design process in various forms (Ylirisku, 2007); it can be collected by designers themselves or by an intermediate researcher. The degree to which the offered information is interpreted can range from raw data collected through an ethnographic approach to design documentaries (e.g. Raijmakers *et al.*, 2006).

Research set-up

In a workshop, three teams were provided with different formats addressing information on real patients' spatial experience in motion. These three teams paired an architect with (1) a geographer, (2) a pedagogue and (3) an anthropologist. The workshop started with an initial brainstorm session focusing on the meaning of being a patient and being wheeled through the hospital. Subsequently, participants were asked to design a lift on the route from the ward to the operation room (OR), based on specific inputs. The task combined designing a moving building element (a lift) for a mobile subject (the patient), supported by a mobile object (the bed).

In a first phase, each of the three teams received a different information format:

1   a written design brief (Table 13.1), mentioning dimensions and other practical information but differing from a traditional brief due to its focus on experiential information (*design-brief team*);
2   a video of a patient's route from the ward to the OR (Figure 13.1), made by a researcher lying in bed and subtitled with the researcher's reflections on embodied perceptions along the route; during the video quotes from real patients appeared when relevant to what was shown (*video team*);
3   a former patient with a background in architecture to talk to (*patient team*).

Table 13.1 Written design brief

Moving designs for moving real people: designing an elevator and the according experience
Hospitals are locations in which a supportive environment is most desired. As a patient you tend to experience these buildings from a rather atypical perspective: lying in a hospital bed. Apart from being atypical, the perspective is also multi-layered.
• The bed as a material object, with its specific accessories, interacts with the built environment around it. Its dimensions and practicalities influence how it is used and experienced by patients.   • However, the bed also has a significant influence on the social interactions you, as a patient, experience while being in the hospital. Unknown people intruding your personal space and relatives and friends keeping a distance are commonplace.   • Moreover, both physical and social interactions are not limited to one location or situation. A hospital bed travels with you through the entire building, as such adding a motional aspect to the hospital experience.
In the elevator all of these and even more elements of the spatial experience of hospital patients are condensed. Therefore, this specific space forms an ultimate challenge to start designing from patients' perspective.

PURPOSE OF THE DESIGN	Obviously an elevator is meant to move people and things up and down in a building. For patients, on the route from a ward to the OR, a hospital elevator is also a transition zone where many actors come together. Due to the limited size of the space, the built environment comes oppressively close to the bed and the person in the bed.

(Continued)

Table 13.1 (Continued)

Moving designs for moving real people: designing an elevator and the according experience	
	As a patient, you are never on your own. A nurse accompanying you pushes your bed in the elevator, and it is him/her that pushes the buttons. To do so (s)he may have to lean over the bed reducing your private space even more. Complete strangers can try to squeeze in or leave the elevator when the patient enters. Patients are wheeled in and out of the small cage of the elevator, but also when the bed stays static, they still move closer to their destination. An elevator, and its influence on the according experience, is thus an example of how a thoughtful design could result in a supportive environment.
REQUIREMENTS FOR THE ELEVATOR	The elevator should be: • a place of transition between the spaces before and after • spacious enough, so it doesn't feel like a cage • made of a warm material, not something that seems to close down on you • pleasant in temperature, so it won't be associated with an oven • easy to operate, without unnecessary wringing of personnel to get to the buttons • supportive in manoeuvring the bed • able to make people feel at ease
PRACTICALITIES	The elevator should be suited to be loaded with: • a stretcher • a hospital bed (for which you need reinforced doorsteps, both at the platform and the cage + a reinforced floor) An elevator for bed transport is approximately $1.4 \times 2.8 \times 2.3$ m (W × D × H). The elevator should be able to stop at each floor (6). Users should be able to get in and out of the elevator at two opposite sites in the longitudinal direction. The operation panel and badge reader should be easily reachable. Each platform should be equipped with operation buttons and a badge reader. Sliding doors are required.
OVERALL	The elevator should make people feel better rather than worse. Some things are obvious: • pleasant lighting • optimal privacy without patients feeling to be neglected • suiting patients' state of mind

Initially, each team consulted only the information format assigned to them, after which they presented the results and the design process. Later the teams made use of all sources of information and could adapt their design resulting in a final presentation focussing on the adaptations. The workshop was audio-recorded, transcribed verbatim and analysed in combination with the design documents generated by the teams.

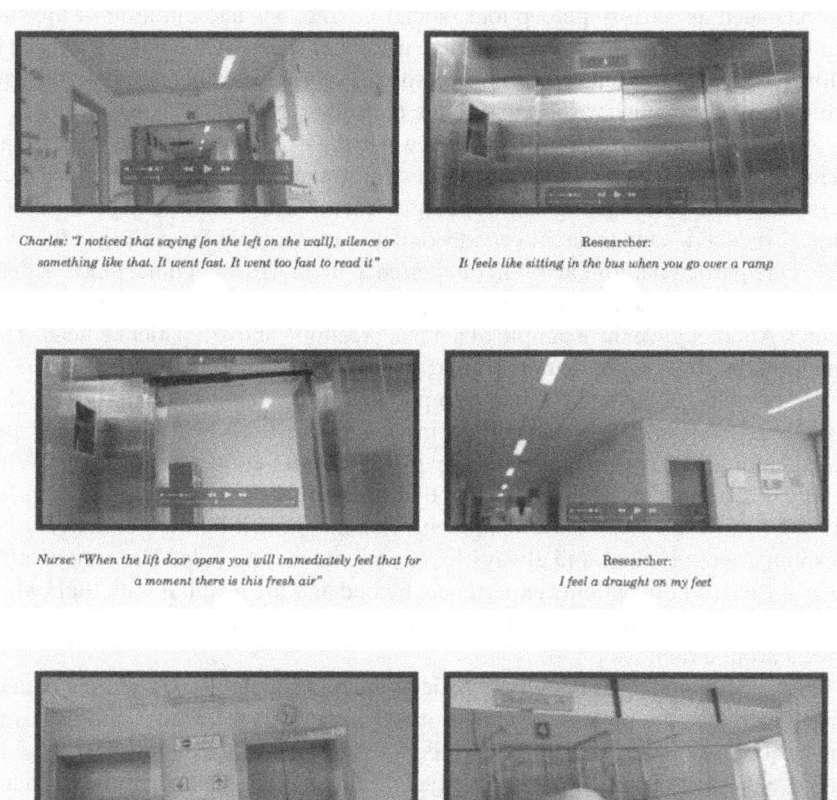

Figure 13.1 Patient's route from ward to operating room.

A final discussion of the design process addressed where, how and why adaptations took place (or not) and considered what role the provided information played in this process. This allowed identifying how the use of the different information formats influenced participants' sense of patients' experience and how this was translated into the design.

Findings

From the brainstorm

During the initial brainstorm session we aimed to achieve a common basis as to whom the different teams would be designing for. Three questions were asked: what does it mean to be a patient (in bed)? What does a hospital mean to you?

What does it mean to be transported through a hospital? Participants mentioned aspects such as sensory perceptions, social interactions and duration of the stay, which was similar to themes identified in an earlier study (Annemans *et al.*, 2011). Most participants had at least some personal experience of being a patient, upon which they drew when responding to the questions.

According to the participants, being a patient in bed means in the first place being under the control of and dependent on a stranger. Due to a patient's changed perspective, a participating anthropologist mentioned that visual perception is largely reduced—and an architect added that indeed mainly the ceiling is then visible. One participant told about her experience in an MRI machine: since patients are unable to see who is present, their experience of others' presence relies more on sound. Another gave the example of a nurse breathing above a patient's head. Thus smells and sounds become more important; and the soundscape can be altered.

Participants described the bed as a patients' only home in the hospital. They live in it, sleep in it, are transported in it, and it is used to transport their personal belongings. As one participant explained from her own experience: "When staff came and changed the blankets, that didn't feel nice, because they made it all new, and I lost my home." Someone else mentioned that the bedcovers in a hospital are really thin, and always leave patients feeling cold. The conversation then shifted to how patients experience the bed and the hospital with their whole body: lying in bed, feeling their own things close to them and the given hospital sheets around them.

Participants reflected upon how patient and bed become one, and how patients thus experience the built environment through the bed. A hospital building is for most people a strange place, disconnected from everything they know or with which they are familiar. Typical sounds emerge from the building, like the 'plong' sounds of the lift. Often the built environment does not seem suitable for the activities taking place. Patients are parked in the hallway to wait. Moreover, many hospitals are said to be ugly, worn down and in desperate need of maintenance. A participant wondered: "if the building is in such a bad shape, then what will they do to me?"

The group concluded that patient, bed and building are connected through transport. Although the hospital bed ties patients to themselves, they are mobile as long as someone is moving them. Being wheeled around compromises patients' sense of orientation: it is difficult for them to know where they are or are being taken: building up a mental map seems almost impossible. Moreover, as patients are often not told where they are being taken, being transported makes them feel like an object being processed rather than a person being taken care of. Movement sometimes happens very suddenly, which can be disturbing. Conversely, one participant recalled that having been in the hospital for a long time, being taken out of her room and wheeled around through the building was also a positive experience.

From the design session

Based on the dimensions mentioned in the experiential design brief, the *design-brief team* started by making a small 3D model of the space they were asked to design (Figure 13.2), a "type of tunnel-shaped elevator" as they called it. From there they made adaptations based on experiential information mentioned in the brief and the ideation during the brainstorm. The design aimed to create a protective corner so people stepping into the lift would not directly bump into the bed. Therefore, one wall of the lift would be curved. They positioned the lift at an outside wall of the building, making the curved wall in glass, so patients would be able to look outside, having a broader perspective and not feeling oppressed. Staff were invited to stand in the additional space generated by the curved surface so they could easily reach the panel to operate the lift. As the architect in the team put it: "through the shape we want to give directions on how to use this space".

The bed itself also was taken into consideration. In the adapted design, beds were equipped with a cover, like a baby pram, with LEDs inside to create a personal ambiance. This would give patients the possibility to withdraw, "like raising the sheet over your head". The design-brief team also listed technical details that would facilitate interaction between patient, bed and building. They mentioned a map of the hospital at the wall so patients would know where they were, a moving platform to ease entry to the lift from the hallway, and an indication of the floors high enough on the wall so a patient could see it from the bed.

Consulting the patient and watching the video made the design-brief team list four adjustments to their design. Changing the window to a screen depicting a landscape could create a better ambience in the lift. It was also a practical decision since this allowed the lift to be situated anywhere in the building, not just at an outside wall. Additional ideas were to provide information technology on the ceiling, or create a daylight ambience. The idea of installing a mirror above the bed, so the patient would be able to observe what was happening when the doors behind the bed opened, was abandoned in dialogue with the patient who thought that seeing yourself as a patient could be frustrating. Finally, also the area in front of the lift was made more pleasant.

From the video, the *video team* (Figure 13.3) started thinking about the lift but "a little bit broader than the lift, the lift as a system that connects the floors". As they put it: "[now] the experience of the space is a little box with a very hard threshold to the surroundings. What if we can see it like a space that's just a continuation of space?".

The video team proposed a "paternoster lift", an existing concept but adapted to the hospital context. Patients would be picked up at the ward and then reside in the lift until they were dropped off at the OR: being wheeled along the hallway would be reduced to a minimum. The concept's disadvantages were transformed into advantages:

> For example it's too slow, but it also holds some good insights, it can be a combination of rooms, so you don't feel oppressed. You just enter and you have a continuous going and maybe the time that you spend, the surplus time that you spend in this loop, is maybe more pleasant than when you have to wait.

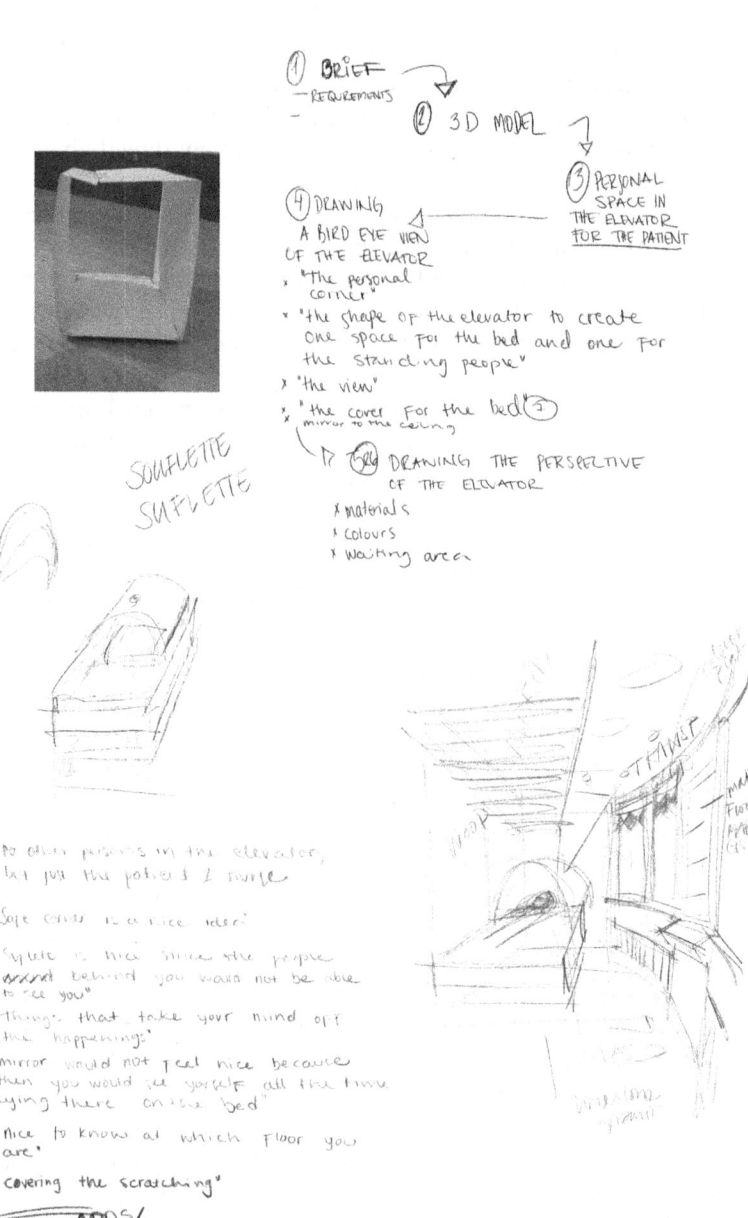

Figure 13.2 Design brief team.

Specific attention was given to entering the lift. The hallway was designed such that the entrance would be smoothed, reducing unnecessary manoeuvring with the bed and bumps at the ridge of the lift door. A connection between patient, caregiver and building was stimulated through the use of mirrors, offering patients a broader perspective than usual when lying down.

In the workshop's second round, the video team continued to work on their "slow lift". They aimed to create a place where you want to be and relax, a continuation of broad spaces rather than an interruption on your route to the OR. This concept was further elaborated in the lift's interior design. A bench would allow the accompanying caregiver to spontaneously sit at eyelevel with the patient lying in the bed. A screen displaying images of nature would provide an interesting focus.

The *patient team* (Figure 13.4) did not have an elaborated design when presenting after the first round. They identified the route as the most important aspect of the patient's story, beginning in her room and moving all the way through to the OR and back: "like a loop she did in the hospital". This loop demanded personalisation, which they wanted to achieve by creating a cover for the bed (just like the design-brief team); however, they quickly abandoned the idea because the patient "was not that into it".

The design idea proposed by the patient team was a personalised path that would unroll for the patient through media architecture. Ideally "the bed would be recognized as your home, and the building would recognize where it would go". The ceiling and walls would then be used to display something visually interesting but not entertainment. The patient had mentioned several times that "reading the magazine she was given or watching TV is frustrating because it's so stressful, the situation is so stressful that this kind of normal entertainment is apparently a little bit banal in that situation". Making use of media architecture would provide patients with something on which to focus, changing the uniform white spaces without interfering with medical procedure.

Consulting the video opened participants' eyes to the awkwardness of the built environment, and the ugliness of some places. This made them look for ways to make the interior more appealing. It made them think more about the hospital interior's materiality: a plain white wallpaper or paint that could be turned into a patient's colour of choice. They further elaborated the ideas of the media architecture, offering patients the opportunity to choose their own theme or colour travelling with them along the corridors, into the lift, and in each room they stayed. In a space like the lift where the bed stood still, the patient could be given additional information, like the estimated time of surgery. The essence of the idea lay in the bed communicating with the building when moving through it, so there would be no need for manoeuvring to call or start the lift. The caregiver would be able to concentrate entirely on the patient.

Design process

The design processes of the three teams diverged considerably. The *design-brief team* started from the most traditional information format, an (experiential)

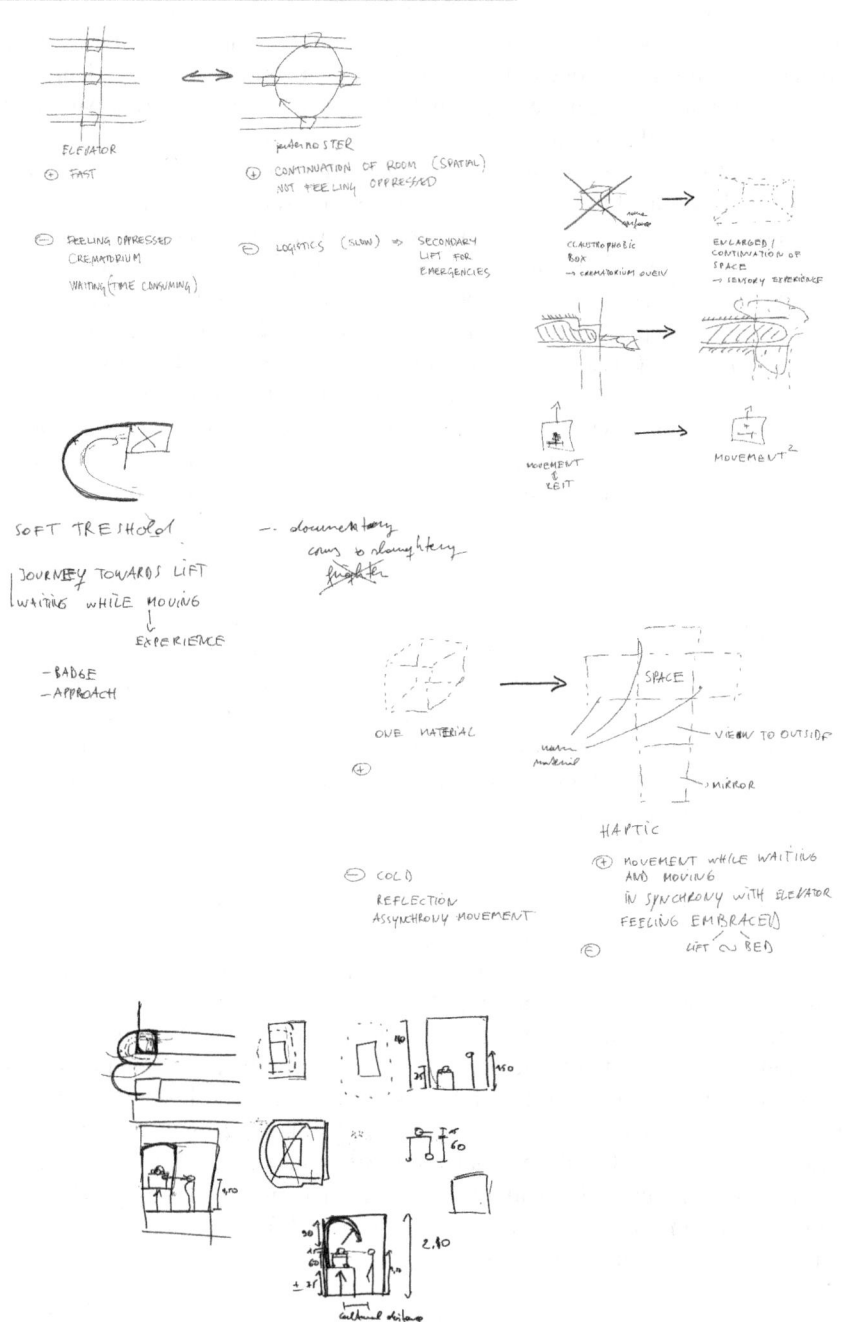

Figure 13.3 Video team.

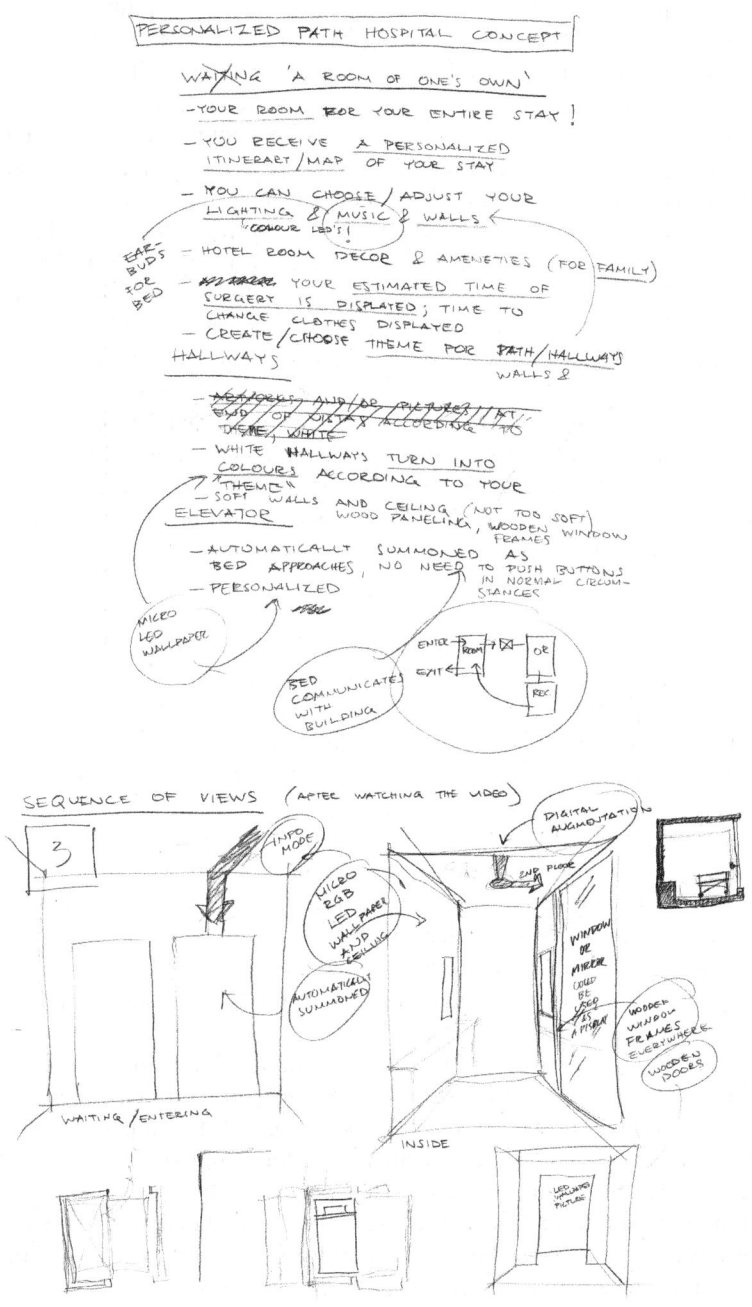

Figure 13.4 Patient team.

design brief. This resulted in a design that kept close to the assignment. The architect in the team found the brief easy to work with and mentioned that he might be biased because he is used to this kind of information. Still, both team members felt supported by all the functional and experiential requirements listed as they provided a framework to keep focused. Some requirements, like the dimensions, also restricted them to a rectangular shape. By stepping back and shifting focus to the patient the team felt empathy through the brainstorm and were able to expand this limiting information. Watching the video and talking to the patient made them scrutinise their own design decisions from a different angle, abandoning previous options: "when we saw the video we started to downsize our ideas. Just because we were confronted with the harsh reality of entering the lift [before, based on the brief] we had to create it in our visionary perception".

The *patient team* indicated how different their design process was from the one explained by the design-brief team. Instead of the spatial focus common in design, dialogue with the patient drew their attention to her story, which took time, also because they reworded her story several times, focusing on different details. This resulted in a much more temporal and experiential focus of the design process rather than a spatial one. As one of the team explained:

> Her experiences happened on a timeline, so we kind of went through this path, [...] to me it unfolded as a path, a spatial path but [...] even though I saw it as a spatial path the process was definitely more verbal. Definitely a lot more verbal than usually, definitely more to do with senses and feelings [...] We didn't get down to the solutions.

Given the importance of the path, or route, it was very hard for the team to focus only on the lift. Emphasising the patient's perspective also made them wonder about the experiences of other actors, like the caregiver. Taking into account all these different perspectives was felt to be limiting the design possibilities. Designing this way was considered rewarding and eye opening, but very time consuming. Watching the video drew their attention to the actual material, spatial reality, they said.

The *video team's* design process was shaped by the combination of a "fresh visual experience", the video with patients' testimonies and the brainstorm session. Seeing the movement of the lift, reading quotes from different patients and being able to situate these in a real environment gave them a broad basis to start designing. Several times they addressed sensory perceptions as a motivation for a design decision, for example: "you could see the ceiling and it made you very sick to see the lines there, so it enhanced the perceptual feeling, so that was one detail that you had to choose the ceiling materials and forms carefully".

Even in the second round of the design, the video team said that they consulted the design brief not at all, and the patient only very briefly, because there were too many items on their to-do list. As one explained:

We didn't exactly come to a design, more to design typologies or things we want to implement. [...] We used this as a communication tool in combination with the video. So it was also time consuming to come to an actual design.

From the discussion

The design processes were shaped by the information formats consulted initially, but regardless of the order in which they were consulted, each had its own merit. As one participant from the patient team explained, her focus shifted from the literal perspective of a patient in a bed over the flow of space, to the patient's story:

> The [patient's] story was a lot more compelling than the main exercise but during the first exercise I thought well, if I'd have to start designing right now, maybe I'd lie down on one of the desks here. But then I watched the video and I thought—what struck me most was the flow of spaces [...] entering the lift was dreadful, but in the whole video it was about really being able to move through spaces. I got that from the video, so I got different things from each thing. Her story obviously didn't have the [dimensions] the brief had. So if I'd really have to do some work, design work, I'd need that.

The value the different teams assigned to the brainstorm session varied. The design-brief team highly valued it: "We had the traditional, even mechanical brief from the client, but this session before, that affected a lot of things".

The design-brief team became aware of their own experience through the brainstorm; and described it as a primary layer, not directly related to the design assignment, but essential for their design. As they said: "We started the process like, let's imagine we were lying down, the perspective you pointed at and how it would feel. We didn't start from the brief". For the other teams the brainstorm session guided the final result to a lesser extent.

Despite its experiential character, the brief could not compete with a real testimony:

> When I started reading the brief, the first sentences were about [the experience of] lying down and being in a weird position. But it was striking, after listening to [the patient] for 45 minutes, [the text of the design brief] felt really flat, obviously.

However, the same person referred to the dimensions in the brief as essential for an actual design. Whereas the brief still had to be analysed and confronted with the ideation from the brainstorm, the video clearly identified the problems of being wheeled in a bed, as such easily allowing the video team to start thinking of solutions: "I had the feeling that we could step over some phases to come to a design. I'm not sure, but we went straight to— [...] nobody had to tell us anymore what the problem was". As mentioned above, it showed the flow of spaces, something that the other sources did not seem to do. "The video really feels like you're walking in someone's shoes" (participant from the design-brief team).

Implications

Developing empathy, becoming connected with the mobile subject and understanding the specific situation one is designing for are indispensable for designers. Here we have demonstrated how understanding of patients' specific situations is influenced by the information formats used to represent and communicate it. The ideation during the brainstorm encouraged designers to reflect on personal experiences (if any) in the given situation. Although the session sensitised those who did not have personal hospital experience, their insight could not go further than imagining.

A mobile perspective (on the environment) is multifaceted. Spatial experience in motion is a complex, layered phenomenon combining sensory perception, social characteristics and time-related aspects (Annemans *et al.*, 2011). Sensory aspects were mentioned, but not yet fully developed in the designs. The designers argued that implementing these aspects would come in a later phase when they started thinking about the materials. The mobile subject is composed of and determined by social relations. In the case of a hospital bed, patient, caregiver and bed are moving together. The perspective from which patients observe their environment while lying in a hospital bed largely determines their interaction with others. In relation to time, all teams addressed the spending of unoccupied time.

Experiencing space while moving is not a linear process. Although different spaces are moved through sequentially in time, various impressions are perceived in parallel. The hospital environment is known for its uniformity. Most wards and corridors look identical: they have plain white, grey or beige walls and suspended ceilings. Many design elements seem to be determined by unspoken hospital procedures. A traditional hospital brief does not challenge designers to think beyond what is known. Further, because textual communication on its own is intrinsically sequential, it seems unsuitable for capturing spatial experience in motion. Although verbal communication with a patient provided designers with a rich, profound insight into her experiences, they appeared to gain insight mainly from a time related perspective, grasping moments, not spaces or flows. Despite the valuable input from the patient and the video, a member of the design-brief team said to the patient:

> It would be different if we could be in the room with you, if we could go through the route, we could sense the environment ourselves. I don't know how to translate it to the design, but still it would be very different […] even watching it from the video, you don't get all the senses.

Conclusion

An analysis of workshop outcomes allows us to formulate recommendations about what should be communicated and how to mobilise design in the hospital. The workshop's aim was to gain insight into how different information formats introducing mobility in design influence architects' design processes. We conclude that in order to create empathy and connect designers into the worlds of those they are designing for, information formats should address designers' background and experiences. In order to attend to mobile perspectives the information format should depict the flow of space in combination with personal stories in a layered, not necessarily linear way. For a nuanced and rich design result, architects need to feel the environment, not from one person's perspective but from several perspectives, including their own.

Despite its experiential character, the design brief fell short of addressing the nuances and details of patients' experiences. Talking with a real patient in this case worked as an eye opener for the designers. However, given that the selected patient was trained as an architect, we can assume that she had a significant advantage in communicating her spatial experience. We cannot expect all patients to be able to do so as eloquently. Although consulting real patients is advisable, fully addressing user experiences can be challenging, even for architects who are strongly committed to such projects (Sanders, 2009).

Whereas video material showed the flow of spaces and thus drew designers' attention to space in motion, the patient's story was obviously personal, inspiring a personal space that moved with the patient. The design solutions point to two distinctly different interpretations of mobile space: an actually moving space in which patients can reside, or a virtual space, moving along with them. The meaning addressed through language, as in dialogue with the patient, differs significantly from the meaning conveyed through embodied use as presented in the video (Clapham, 2011).

The workshop's results suggest, however, that providing designers with different information formats makes them question the environment as a static given. Combining words and images or, even better, moving images, seems promising but obviously lacks the interactive element of a conversation. In line with work by McGinley and Dong (2011), we can conclude that the challenge of introducing insights regarding spatial experience in motion in architects' design process relates both to the content and format of the information. The content should be as close as possible to raw data, reflecting patients' own testimonies. The information format should be able to convey a nuanced image of the research findings and preferably facilitate interaction. Passing on video material supported by an extended narrative, with the opportunity to consult more information than what is initially provided in the flow of images, seems promising in supporting design for mobility in the hospital context. Further research is needed to develop a format that can actually support this aim.

Acknowledgements

Margo Annemans' research is funded by a PhD grant of the Baekeland programme of the Agency for Innovation by Science and Technology in Flanders (IWT-Vlaanderen), which gives researchers the opportunity to complete a PhD in close collaboration with industry, in this case osar architects nv. Ann Heylighen received support from the European Research Council under the European Community's Seventh Framework Programme (FP7/2007–2013)/ERC grant agreements no. 201673 and no. 335002. The authors thank the organisation of the 6th Annual Symposium of Architectural Research 2014 for the organisation of the workshop, the participants for sharing their time and insights and the patient for her generous input and cooperation.

References

Annemans, M., 2015. *The experience of lying: informing the design of hospital architecture on patients' spatial experience in motion* (PhD manuscript). KU Leuven, Faculty of Engineering Science, Department of Architecture, Leuven.

Annemans, M., Van Audenhove, C., Vermolen, H. and Heylighen, A., 2011. Lying architecture: experiencing space from a hospital bed. *Proceedings of Well-being 2011: the first international conference exploring the multi-dimensions of well-being.* Birmingham: Birmingham City University and the RIBA.

Annemans, M., Van Audenhove, C., Vermolen, H. and Heylighen, A., 2012. Hospital reality from a lying perspective: exploring a sensory research approach. In *Designing inclusive systems designing inclusion for real-world applications.* Ed. P. Langdon, J. Clarkson, P. Robinson, J. Lazar and A. Heylighen. London: Springer, pp. 3–12.

Annemans, M., Karanastasi, E. and Heylighen, A., 2014. From designing for the patient to designing for a person. In *Inclusive designing: joining usability, accessibility, and inclusion.* Ed. P. Langdon, J. Lazar, A. Heylighen and H. Dong. London: Springer, pp. 189–200. doi:10.1007/978-3-319-05095-9_12.

Bogers, T., van Meel, J.J. and van der Voordt, T.J.M., 2008. Architects about briefing: recommendations to improve communication between clients and architects. *Facilities* 26: 109–116. doi:10.1108/02632770810849454.

Cassim, J., 2010. "It's not what you do, it's the way that you do it". In *Proceedings of the 3rd International Association of Universal Design. International Association for Universal Design.* Ed. C. Stephanidis. Hamamatsu, pp. 36–45.

Clapham, D., 2011. The embodied use of the material home: an affordance approach. *Hous. Theory Soc.* 28: 360–376. doi:10.1080/14036096.2011.564444.

Elf, M. and Malmqvist, I., 2009. An audit of the content and quality in briefs for Swedish healthcare spaces. *J. Facil. Manag.* 7: 198–211. doi:10.1108/14725960910971478.

Elf, M., Svedbo Engström, M. and Wijk, H., 2012. An assessment of briefs used for designing healthcare environments: a survey in Sweden. *Constr. Manag. Econ.* 30: pp. 835–844. doi:10.1080/01446193.2012.702917.

Howard, Z. and Somerville, M.M., 2014. A comparative study of two design charrettes: implications for codesign and participatory action research. *CoDesign* 10: 46–62. doi:10.1080/15710882.2014.881883.

Ingold, T., 2011. *Being alive: essays on movement, knowledge and description.* London. Routledge.

Kouprie, M. and Visser, F.S., 2009. A framework for empathy in design: stepping into and out of the user's life. *J. Eng. Des.* 20: 437–448. doi:10.1080/09544820902875033.

Lofthouse, V., 2006. Ecodesign tools for designers: defining the requirements. *J. Clean. Prod.* 14: 1386–1395. doi:10.1016/j.jclepro.2005.11.013.

McGinley, C. and Dong, H., 2011. Designing with information and empathy: delivering human information to designers. *Des. J.* 4,: 187–206.

Raijmakers, B., Gaver, W.W. and Bishay, J., 2006. *Design documentaries: inspiring design research through documentary film.* ACM Press, p. 229. doi:10.1145/1142405.1142441.

Restrepo, J., 2004. *Information processing in design.* Delft: Delft University Press.

Sanders, L., 2009. Exploring co-creation on a large scale. In *Designing for, with, and from user experience proceedings.* Ed. P.J. Stappers. Delft: StudioLab Press, pp. 10–26.

Van der Linden, V., Annemans, M. and Heylighen, A., 2016. Architects' approaches to healing environment in designing a Maggie's Cancer Caring Centre. *Des. J.* 19: 511–533.

Van Rijn, H., Sleeswijk Visser, F., Stappers, P.J. and Özakar, A.D., 2011. Achieving empathy with users: the effects of different sources of information. *CoDesign* 7: 65–77. doi:10.1080/15710882.2011.609889.

Ylirisku, S., 2007. *Designing with video: focusing the user-centred design process.* London: Springer.

14 Border crossings

Exploring artefacts of mobility with blind and visually impaired users

Jayne Jeffries and Peter Wright

Introduction

This chapter brings together mobilities research in human geography with the field of human–computer interaction (HCI), first, to examine the changing relationships between 'designers' and 'users';[1] and second, to understand the shift from objectified knowledge production to working with users in more collaborative and meaningful ways. We address the histories of design thinking, from human factors engineering to the more recent shift to experience-centred design (Wright and McCarthy, 2008). Using experience as a catalyst we explore the borders that exist between 'designers' and 'users', showing that movements and connections across disciplinary borders enact more fluid and transformative designer–user relationships. We argue that feminist and participatory geographies are well placed to understand how borders become blurred, shifting to accommodate and connect the perspectives, skills and experiences of designer-researchers and participant-users as they negotiate the design process.

Despite a long history of disabling practices across disciplinary boundaries (Chouinard, 1997; Imrie, 2001; Worth, 2013), we use border crossings to explore the potential of inclusive design practices in HCI and feminist geographies. While behavioural approaches to geographical and psychological analyses have been critiqued for objectifying users and (re)producing exclusionary design practices, experience-centred design draws parallels with feminist and participatory approaches, seeking to integrate and connect the voices and experiences of users in order to create more meaningful change. By adopting a methodological approach that seeks to shift interactions from designing 'for' to designing 'with' users, we show that citizenship involves the negotiation of power relations. In the process of mobilising design we use the concept of citizenship as a practice that "disrupt(s) already existing norms and practices" (Spinney *et al.,* 2015, 328).

The discussion below draws upon a wider interdisciplinary project: *MyPlace: Mobility and Place for the Age Friendly City Environment*, which worked with a range of citizens to collect and document people's experiences of their changing mobilities.[2] The case study, which forms the focus of this chapter, is developed from the first author's work with users at Henshaws, a charitable organisation for blind and visually impaired users. The research sought to draw attention to 'users', to whom we refer as citizens who are often marginalised or excluded from the design process, as well as to understand and reflect on the role of 'designers'. Using a qualitative, participatory approach, this chapter draws upon research between March and November 2015, focusing on designer–user interactions. The analysis here is distinctive in extending beyond considerations of how 'different' bodies move through and inhabit the built environment (Evans and Colls, 2012) to foreground the ways in which visually impaired users interact with artefacts of mobility.

Border crossings

Borders provide an important conceptual framework to explore the histories of mobilities research, from work addressing cross-border migration patterns to borders as a mechanism that holds back the movement of people, often containing or controlling citizens. This chapter shifts focus from the fixed and bounded notion of mobility and movements across the globe (Urry, 2002) to the opening up of more fluid borders. We argue that citizenship is an important mechanism for understanding the borders between (dis)abling design practices of inclusion and exclusion. We illustrate the blurring of theoretical and disciplinary borders, first, by exploring the role of design thinking in HCI and second, by examining geographical approaches to citizenship, impairment and mobilities.

Design and human–computer interaction

In this section we explore the history of design thinking, examining the changing relationships between designers and users and the shift to more fluid borders and territories between disciplines and professional practices. We anchor our discussions in the context of HCI, situating the practices of design professionals and unpicking the fundamental approaches to 'design' and to the changing role of the 'user'.

The field of HCI emerged around the late 1980s as a distinct discipline, and as digital technology migrated from the shop floor and office into people's homes and into their pockets. HCI challenged more traditional conceptions of the human as an operator in a larger technical system, one that defines categorical 'norms' of optimal 'human' performance by focusing on the behavioural capabilities of individuals in relation to 'non-human' components of a system. These (non-)human relationships are prevalent in ergonomic and human factors engineering, which reinforces designing 'for' and objectifying users through psychological analysis of human behaviour. In this context, design would seek to restore the capabilities of a vision-impaired person by (re)allocating the functions of the visual channel to (for example) the tactile. In contrast, this chapter emphasises that sight loss does not necessarily exclude people from visual modes of communication, a factor overlooked by earlier design practices.

The turn away from human factors and ergonomics was accompanied by new modes of user-centred research, emphasising the need for designers to 'know their users' and to understand the context of their work (Norman and Draper, 1986). Sites of human–computer 'interaction' design were located in an understanding of the user through research exploring social phenomenological (Dourish, 2001), ethnographic and ethno-methodological (Suchman 1987) methods. Technology use as a social practice both reflected and opened up new technology design spaces, providing a critique of earlier cognitivist approaches to user-centred design. These new design spaces are illustrative of the 'turn to experience' in HCI, which is discussed by McCarthy and Wright (2004).

Wright and McCarthy (2010) argue that social practice accounts of HCI in contexts as varied as work, home, education and leisure understate 'felt life'. Further, in order to do justice to the wide range of influences that technology has in our lives, there is a need to make sense of the relationship between user and technology in terms of the felt or emotional quality of action and interaction with others and with technology. Taking inspiration from the work of Bakhtin and Dewey, Wright and McCarthy (2010) develop a pragmatist account of *technology as experience* in which the emotional-volitional and creative aspects of people's relations with technology are foregrounded. Here the meaning and value of tech-nology to a person is not fixed but changes through processes of appropriation and assimilation where both technology and person are connected and transformed. Technology is put to uses never envisaged by designers, creating new possibilities and new meanings, while simultaneously people find new identities and new ways of being through the affordances offered by technology.

A relational conception of HCI also challenges taken-for-granted concep-tions of what it means to 'know' the user and the hierarchical relations between designer-researcher and user that knowing can sometimes imply. Wright and McCarthy (2010) develop a conception of designer-researcher and user as differ-ently placed experts who share experiences in order to try and comprehend what it is like to be 'the other'. This dialogical approach acknowledges differences in the expertise that designers and users bring to the design process but seeks through mutual appreciation of the potential offered by 'other' to create design possibili-ties that neither side alone could have anticipated (Wright and McCarthy 2008, Wallace *et al.*, 2013).

This perspective is key to understanding the shift from objectified knowledge production towards the changing role of the user and the active role of citizenship and belonging. In this case study we seek to explore assumptions associated with the 'visual' and visually impaired users' skills using feminist dialogue.

Citizenship and mobilities

In this section we continue to examine citizenship as a practice, arguing that the tendency to marginalise certain subject positions, such as those on the edges or margins of society has shaped feminist analyses of gender, sexuality, ethnicity and race in the context of landscapes of belonging (Rose, 1993; Tolia-Kelly, 2006). A sense of belonging, or lack of, may also suggest that certain citizens are made to feel different, in a variety of social situations and interactions, including people who are visually impaired.

A number of human geographers argue that these disabling practices often make people feel 'out of place' (Kitchin, 1998) in education (Holt, 2003), the workplace (Chouinard and Grant, 1995) and everyday social life (Parr, 2008). Chouinard (2009, 108) argues that feminist geographers are well placed to understand the way "citizens continue to be denied rights and entitlements in the neoliberal societies of the early twenty-first century", drawing attention to the role that practices of inclusion and exclusion play in establishing borders. There is a collective feeling that citizenship plays out through power relations, but that "disabling forms of citizenship have received little attention in the geographic literature" (ibid., 111).

More critical work that engages with marginalised citizens is needed to explore the role of belonging and active engagement in the design process, as gaps still remain in the way so-called "distinctive subject positions" (Cresswell and Merriman, 2011, 9) are represented in mobilities research. Hannam *et al.* (2006, 3–4) state that "[i]t is not a question of privileging a 'mobile subjectivity', but rather a tracking of the power and politics of discourses and practices of mobility in creating both movement and stasis". However, when thinking about impaired users there has largely been a privileging of the micro-geographies of the (able) body[3] within the mobilities paradigm (Cresswell, 2011).

There are two strands of research addressing the movement of sensory impaired populations, first, an understanding of behavioural approaches in relation to users mobilities. For example, Golledge's (1993) work on spatial analyses of wayfinding for blind and visually impaired users shares in the earlier design practices of human factors engineering, adopting a tendency to exclude 'users' subjectivities in favour of objectification. In a similar vein, Imrie (2001) has drawn attention to exclusionary practices in the design of urban environments, critiquing 'shared space' for continuing to position visually impaired users at a disadvantage. Imrie (2012, 2265) castigates urban policy makers for creating "[a]uto-disabling spaces [that] are reflective of the oppressive social norms, beliefs, and values that seek to prioritise movement and fluidity of the mobile body [. . .] with the flows and rhythms of the motor vehicles".

Second, there is an interest in qualitative accounts of first-hand experiences, including Worth's (2013) work on the socio-spatial relationship between bodies, power and visually impaired young people's daily movements. This research highlights strategies such as "wearing headphones while travelling a familiar route on the bus or train" (ibid., 583) to avoid unwelcoming questioning, using "mobility to try and negotiate an ableist society" (ibid., 576). However, when working as a sighted guide in the Lake District, Macpherson (2009) sought to understand how blind and visually impaired walkers find their way by 'feeling through the feet'. Significantly, her work shows that "bodies and their differences come about through their interactions with the world" (Macpherson, 2010, 4). It is in this context that we argue that blind and visually impaired users' mobilities change by interacting with artefacts or technologies—a dimension largely absent from the mobilities literature.

Below, we analyse the importance of 'locations' in mobilising design, first, by examining the organisational setting where the first author held weekly workshops 'in place'; and second, by exploring user-led design practices interactions with artefacts of mobility that unfolded whilst 'on the move' with visually impaired users.

Case study: Henshaws

This section reflects upon the design process that developed when working with users at Henshaws, a charitable organisation in Newcastle-upon-Tyne that provides "expert support, advice and training to anyone affected by sight loss" (Henshaws 2015). The first author established a long-term relationship with Henshaws over a ten-year period, building trust and familiarity with staff and volunteers, enabling her to gain access to a new group of blind and visually impaired users. Nine service users, including one blind user and eight visually impaired users[4] were involved in the research from March to November 2015.

Methodology

We use inventive mobile methods (Lury and Wakeford, 2012) to describe the informal and unintentional basis of our approach. This incorporates what Cresswell (2011) refers to as mobile ethnographies, including a range of qualitative methods that engaged service users during periods of time staying 'in place' at Henshaws' office and 'on the move' travelling in and around Newcastle-upon-Tyne.

The first author attended pre- and self-organised activities such as bowling, rock climbing, raft building and cycling, as well as a number of trips to the coast and a full day at a local open-air museum. Empirical data from qualitative methods is taken from the first author's research diary, facilitated conversations, a user-led guided walk, visual methods and one-to-one interviews. First, we examine the importance of designer–user relationships 'in place'; and second, we look at the way in which mobile artefacts become subtly embedded in blind and visually impaired users' accounts of being 'on the move'.

In place

Henshaws' work with service users to achieve their own '*Pathways to Independence*' encouraging and supporting users to make their own decisions. This collaboration 'with' service users is a fundamental aspect of the feminist ethos encompassed in this case study. The first author combined her skills as a participatory and qualitative researcher with the dialogical approach outlined in Wright and McCarthy's (2010) 'experience-centred design', aiming to connect with and understand what it is like to be the 'other'[5] by acknowledging differences in the expertise that designers and users bring to the design process. We explore two lessons learned when engaging in dialogue with visually impaired users.

The support offered by Henshaws was explained by the Community Services Manager:

> It's all different, because it's based on each individual's different need [. . .] So when we get referrals, I often say to people, "Right, we'll come and visit you at home, and then you're in the comfort of your own surroundings. You can have whatever family there you want". Then we can sit and talk about what their eye diagnosis is, what the long-term complications will be. We can look at how we're going to support them as an individual, what they want to gain out of life, what courses they need to go on. We then look at their mobility. We then work with the family to say what support needs we're going to put in place. I find that in their own home, they're more relaxed.
>
> (Interview, October 2015)

The quotation highlights the importance of being 'in place' when establishing service users '*eye diagnosis*', what they want to '*gain out of life*', as well as their specific '*mobility*'. In this example the home is an important place for initiating the referral process, which often marks the beginning of a new journey for blind and visually impaired users. However, in the context of mobilising design we discuss what it means to 'know users', discussing the silences that unfold when working together 'with' users 'in place'.

Educational background

The first author began to understand these silences at the community centre where users had been attending educational classes together since September 2014. The participant-users attended a range of English, Maths and IT classes, which were provided by qualified teachers from the Local Education Authority. A row of eight computers occupying two internal walls, a separate cluster of tables with sufficient room for ten people to be seated and a large plasma television screen were placed at one end of the open plan office; the Community Services staff worked at the other end of the room. It was here that the designer–user intersection developed, where one researcher-designer (the first author) and ten participant-users met 'in place' for the first time to learn about and from experience.

As a feminist researcher it is important to understand silences by engaging in dialogue with users, including interactions that allow time to explore personal accounts of health and (visual) impairment. Insights often came from spending time with users, who brought issues into conversation while eating lunch in pairs or smaller groups, rather than during more formal activity sessions. For example, the first author recalls:

> I went to eat my lunch with [four of] the other participants who were sitting around the table, two young girls in their twenties and two men, one I would later find out was a volunteer and the oldest member of the group as he'd like to jibe, the second, [. . .] in his thirties. I chatted informally while I ate my

lunch and it was at this point that one of the girls started to tell me [. . .] about failing all of her exams at school, as she didn't receive any help or extra support with her work. She was born with a VI, although this wasn't recognised until a much later date.

(Research Diary, March 2015)

The experience of being ignored in a school classroom context subsequently shaped users' confidence, including the way they responded to our research activities 'in place'.

Assumptions about the 'visual' / 'other'

Qualitative and participatory methods were employed to gain a deeper understanding of a small number of visually impaired users and their mobilities, including ice-breakers; short, focused activities designed to engage participant-users and designer-researchers in conversation. The aim was to elicit a more informal and relaxed environment by prompting users and designers to share experiences of working with Henshaws. These activities provided opportunities to explore the complexities of visually impaired users' needs, including the role of visual methods, the challenges of audio (recordings) and inherent assumptions about the 'visual' / 'other'.

In earlier design practices, such as in human factors engineering, the relationship between visual modes of communication and sight loss were often taken for granted. However, we argue that there is a danger that other senses are overlooked when assuming a tactile narrative and ignoring the visual in the design process, as illustrated by the following visual methods.

In the first personal profiles activity, users were asked to take a passport-style photograph using a Polaroid camera and to provide other personal details on the A3 template. However, visually impaired users were often reluctant to be the first person to start writing in blank boxes on the template, or taking photographs, a task that one of the younger women in her twenties decided to complete, encouraging other users by taking their photographs as well.

Users completed a second exercise—body mapping— in pairs, initially taking turns to draw around the head and hands of their partner on large (A1) sheets of paper. The users then responded to the question: '*What does independence mean to you?*',[6] subsequently filling the paper with information which was then shared with the larger group (if agreed to by users). Figure 14.1 is an example of one exercise, showing the importance of being '*independent*', '*getting out*', '*getting fit*', '*learning new skills*' and '*Henshaws' training*'.

The use of audio recordings as a substitute for traditional hand-written or typed entries has been used to elicit qualitative diaries by researchers working with visually impaired users (Worth, 2009). However, in the case study we received contradictory results when three participant-users wanted to 'write' rather than (voice)'record' their diaries, stating collectively that they didn't like the sound of their own voices. However, one of the silences that unfolded from participant-users' response to this method included their reluctance to 'write'[7], which came through a facilitated group

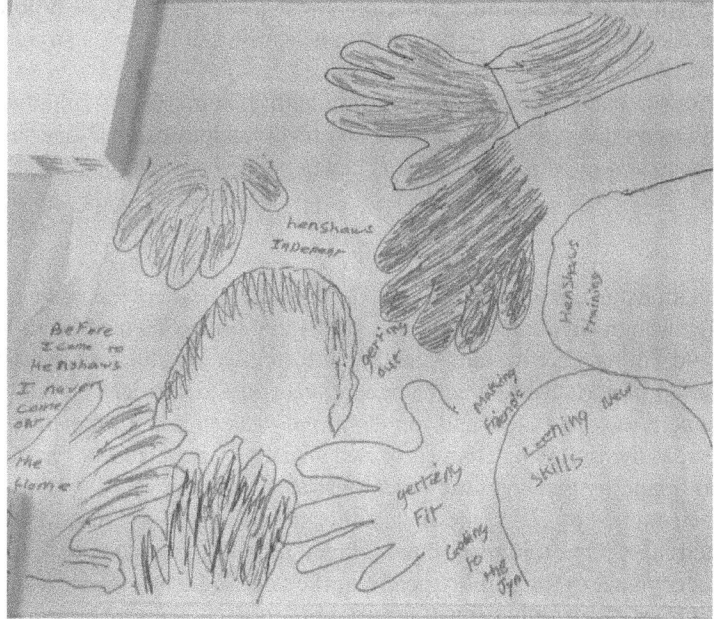

Figure 14.1 Body mapping, 'in place' activity, 2015.

conversation at a later date. While the users agreed to electronically recording this conversation, an act that often gave rise to uncertainty during the activities 'in place', we spoke in detail about 'who' would write the 'steps to planning a journey':

J: Got any takers for writing?
G: No.
P: No. Not me. My writing's terrible.
G: No. Mine is.
J: It doesn't matter about the writing
P: Gary, go on. You do it and I'll tell you how to spell.
J: We can write it and then, if you want, we can re-copy it if you're not happy with it being clear. So, I've just brought some paper.
P: No. Gary's writing and I'll tell him what to write.
G: I cannot write. I tell you.
P: My writing's terrible. You're not getting me writing. Adrian?
G: Adrian?
A: No.
[. . .]
J: Do you want to tell me and I can write them down?
G: Yes.
P: You ask all the questions. We'll give you the answers.

Later in the research, the Community Services Manager reflected: "95% of people with medium to major learning difficulties [. . .] have got a sight problem, but it's never identified or needed, because their learning difficulty is so major" (Interview, October 2015).

However, this in itself is not an easy topic to identify, or one that is volunteered or 'voiced' by users demonstrating that literacy levels shape designer–user interactions and the extent to which connections can be made between the two.

On the move

The design relationship continued to develop 'on the move', where topics emerged intermittently through encounters at the bus stop and journeys on the bus, as well as walking and sitting together and chatting informally. As this demonstrates, journeys were an important site where connections could be made. The relationship between activities 'in place' and 'on the move' was an important part of the design process, as the following analysis of two visually impaired users' 'in place' reflections on 'planning a new journey' demonstrates.

During the summer of 2015, the first author attended pre-organised trips to an outdoor activity centre, a journey that included travelling on a bus with users, as well as a short walk to the activity centre at the end of the bus journey. It was through the design process that fixed and mobile artefacts were revealed as an important aspect in sensing the city. The analysis explores 'knowing place(s)', using quotations to examine the (un)familiarity of places and the relationship users have with artefacts of mobility.

Fixed and mobile artefacts

Philippa reflected on a journey we made together for the first time as a group:

> Well, you know when we went to [the activity centre]? I was a bit iffy walking, for the kerbs and the steps, because I didn't know the area, you see. That's why I got hold of somebody [. . .] Well, I nearly fell twice [. . .] I tripped on the kerb, you see, because I don't know the area.
>
> (Interview, September 2015)

When asked about the steps she would take when planning a new journey, she continues:

> Well, I would have to take someone with me, because I don't know the area and I'm frightened in case I miss the kerb [. . .] Once I get to know the route and I know there are kerbs and stuff, I'm fine, but otherwise I hate going to places on my own the first time.
>
> (Interview, September 2015)

These quotations illustrate the importance of 'fixed' artefacts of mobility, which shape the way users navigate the urban environment. Philippa repeatedly refers to physical objects such as 'kerbs' and 'steps', showing that multiple visits are required to learn the location of these objects and to prevent feeling uncomfortable in an unknown area or being 'frightened' about the possibility of falling or tripping. Philippa continues to discuss the initial journey to the pre-organised activities at the outdoor centre, telling the other users:

P: I was giving Jayne lessons on the crossing.
J: You were giving me lessons. It was like a test. She said, "Jayne, what's this for? Jayne, get hold of the cone. Do you know?" I was like, "I do know about the cone".
P: [*Laughter*] What's that thing on the path for? The bumps?

Gary provided a more in-depth account of 'cones' and 'bumps':

G: Yes. He [the trainer] tells me to put the tape on my cane, on my sticks [. . .]
J: Did you tell him where to go?
G: Yes. Into the town and stuff like that. Round the town. He took me down to feel the bubbles on the crossing and stuff [. . .] He said, "You should be feeling them with your sticks". Because when you get your stick on one of them bubbles. He said, "Can you feel it?". I said, "Yes". He said, "All your crossings have the cones".
J: So did you know about them before you went out with [the trainer]?
G: No [. . .] He [the trainer] said, "Always press the button and feel it because sometimes the green man does not flash". I said to him, "[S]o what happens if you are crossing the road and that just stops beeping?". He said, "You are still entitled to cross that road" [. . .] Once you're on that crossing he said, "It doesn't matter if you are on crutches, sticks or crawling. Those drivers have got to give you time to get across that road".
J: Yes. Did you feel nervous about that?
G: Yes. Because before I got my sticks, I collapsed on a busy main road.

(Interview, October 2015)

The extract highlights Gary's knowledge of a number of existing artefacts of mobility, including fixed and mobile objects that he interacts with during mobility training. Gary refers to "feeling the bubbles . . . with [his] sticks", indicating the presence of tactile or blister paving, colour contrasted red floor tiles with raised concrete or metal "bumps" (Philippa). These tactile surfaces are located at intersections where there is a change in height or road use, such as dropped kerbs at pedestrian crossings.

However, other aspects of micro-design are less visible. For example, Gary's trainer tells him, "All your crossings have the cones", small funnel-shaped objects that are located on the underside of the Pedestrian Demand Unit (PDU). These cones are designed 'for' blind and visually impaired users, replacing an audible beeping sound with a tactile rotating cone (which is hand-held by a user at a crossing) to indicate that it is safe to cross. By aiming to eliminate simultaneous beeping sounds at co-located crossings, they are designed to provide a less confusing environment for visually impaired users. Yet knowledge of these fixed artefacts is unevenly spread amongst blind and visually impaired users, who often only learn about the location of tactile cones during mobility training. In Gary's circumstances this takes place after he 'was knocked down, [having] stepped out in front of a pedal bike" (Gary, Interview, October 2015). Gary's mobility trainer leads him 'into' and 'round the town' (the city centre), feeling urban fixtures such as the 'bubbles' 'through [his] feet' (Macpherson, 2009) and 'sticks'. "He [the trainer] tells me to put the tape on my cane, on my sticks" (Gary).

While Gary uses his crutches on a daily basis, Philippa's use of her long cane is different:

> I could use [it] at times when crossing the road [. . .] [I] can feel what the road is like, for obstacles and uneven surfaces [. . .] [B]ut I don't want my neigh-bours to know, because they are nosey.
>
> (Group Conversation, 2015)

The limited use of her cane also illustrates the importance of 'confidence', a key theme that was explored during body mapping and outlined in the earlier section 'in place'.

The design relationship unfolded as the research continued and it was impor-tant that users determined where we went and why, shaping the self-organised activities. It was Gary who initiated the user-led guided walk to review pedestrian crossings, bringing the issue of a broken crossing to the attention of the Services Manager when arriving at one of the sessions 'in place' during the earlier stages of the research: "That still hasn't been fixed yet. Still only working on one side [. . .] And I stuck that repair in over a year ago now" (Gary). The user-led accompa-nied walk to explore rotating cones, the existing artefacts of mobility, highlights the importance of finding 'things' and the role of design. As Suchman (2011, 1) states, "design needs to acknowledge the specificities of its place, to locate itself as one (albeit multiple) figure and practice of transformation". In this example visually impaired users find new identities and new ways of being through the affordances offered by technology, as fixed and mobile artefacts of mobility change the relationship users have in navigating the built environment.

Conclusion

We return to the designer–user relationship to understand what has been learned through the process of 'experience-centred design', which we argue opens up the possibilities for the borders claimed by designers and users to be shifted and connections to be made. In the case study, blind and visually impaired users bring their own accounts into dialogue with a designer-researcher, including the (un)familiarity of place(s) and their own learning of sensory environments. We explore three key contributions.

First, designers are often interested in re-creating ethnographies to understand users' experiences, through simulated walks or when designers try to 'be' the users. However, these practices impose hierarchical relations that more empowering approaches, such as experience-centred design aim to overcome. These are implicated in Norman and Draper's (1986) conception of 'knowing users', which we have addressed by prioritising visually impaired users' own experiences of their mobilities. Using feminist dialogue we were able to explore the complexities of border crossings, learning about specific silences 'in place', including users' educational background, confidence and assumptions about the 'visual' / 'other'. We argue that both designers and users transform the product and process together merging different skills and experiences through a mutual relationship. This sits in contrast to an approach where a sighted 'designer' is used as a substitute because they now understand 'what it is like to be visually impaired'.

Second, there is a history of designing *for* sensory and physically impaired users, often through rehabilitative practices, which illustrates the relationship between bodies and designed artefacts as an extension of the body, from splints, glasses, cane(s), wheelchairs and assistive aids to technologies to enhance the senses. However, our case study has revealed the importance of ethnographies of place(s), which developed by travelling *with* and being guided by users. In experience-centred approaches, the visually impaired user does not become the object of design, or a source of information to inform design, but a teacher for the designer, which was illustrated during Gary's accompanied walk to explore rotating tactile cones at pedestrian crossings.

Finally, using border crossings we gained an understanding of the relevance of fixed and mobile artefacts of mobility whilst 'on the move'. It is through these artefacts that different visually impaired users learn to understand place, from rotating cones and tactile pavements to canes and crutches. Flowing between stasis and movement as they traverse the city, users knowledge of the infrastructure of place, highlights the role of artefacts within networks and constructs their relative (im) mobility. A shift towards the "networks of people, ideas and things moving, rather than in inhabitation of a shared space such as a region or nation state" (Cresswell, 2011, 551) acknowledges the associated material 'things' that enable or potentially limit people's movements in the built environment. However, objects or artefacts may also be adornments or signifiers of in/exclusion and marginalisation, including failed infrastructures and broken artefacts of mobility located at Pedestrian Demand Units.

Notes

1 In the remainder of this chapter we use the terms 'designer(s)' and 'user(s)', as well as designer-researcher(s) and participant-user(s) to highlight the ambiguities surrounding these categories. We argue that the approaches adopted in experience-centred design and feminist and participatory geographies work to problematise notions of fixed subject positions, subsequently blurring borders between contributors in the (research) design process.
2 MyPlace: Mobility and Place for the Age Friendly City Environment. EPSRC Grant No: EP/K037366/1. See http://www.myplace.ac.uk for more information about the case studies.

3 See Gaete-Reyes (2015) for a paper on the way productive bodies are privileged over disabled women's 'rights' as citizens.
4 The first session consisted of ten people, one blind user (a man in his late forties) and nine visually impaired users, including four women (two in their twenties and two in their forties) and five men (one in his mid-thirties, three in their forties and one in his fifties). Two users were also volunteers; one volunteer continued in subsequent sessions.
5 It is important to clarify that the 'other' being addressed here refers to knowledge and understanding that the designer-researcher, who is not visually impaired and the participant-users who are blind or visually impaired bring to the design process. The aim is not to set up binaries, or to label the participant-users as 'different', instead the 'other' addressed here accepts that each person contributes to the design process and that knowledge passes back and forth between designers and users.
6 The question was chosen as it relates to the overarching aim of the organisation and the support offered to users.
7 Although this was not an issue in the body mapping exercise, which was completed as a group.

References

Chouinard, V. 1997. Making space for disabling differences: challenging ableist geographies. *Environment and Planning D*, 15: 379–387.

Chouinard, V. 2009. Citizenship. In: Kitchin, R. and Thrift, N. (Eds), *International Encyclopedia of Human Geography*, vol. 7. Oxford: Elsevier, pp. 107–112.

Chouinard, V. and Grant, A. 1995. On being not even anywhere near 'the project': ways of putting ourselves in the picture. *Antipode*, 27: 137–166.

Cresswell, T. and Merriman, P. 2011. *Geographies of mobilities: practices, spaces, subjects.* Farnham: Ashgate.

Cresswell, T. 2011. Mobilities II still. *Progress in Human Geography*, 36: 645–653.

Dourish, P. 2001. *Where the action is: the foundations of embodied interaction.* Cambridge, MA: MIT Press.

Evans, B. and Colls, R. 2012. Fat bodies walking? Unpublished working paper. In Andrews, G. J., Hall, E., Evans, B. and Colls, R. Moving beyond walkability: Three commentaries on the critical potential of health geography. *Social Science & Medicine*, 75: 1925–1932.

Gaete-Reyes, M. 2015. Citizenship and the embodied practice of wheelchair use. *Geoforum*, 64: 351–361.

Golledge, R.G. 1993. Geography and the disabled: a survey with special reference to vision impaired and blind populations. *Transactions of the Institute of British Geographers*, 18: 63–85.

Hannam, K., Sheller, M. and Urry, J. 2006. Editorial: Mobilities, immobilities and moorings. *Mobilities*, 1: 1–22.

Henshaws 2015: https://www.henshaws.org.uk, accessed 23 October 2015.

Holt, L. 2003. (Dis)abling children in primary school micro-spaces: geographies of inclusion and exclusion. *Health & Place*, 9: 119–128.

Imrie, R. 2001. Barriered and bounded places and the spatialities of disability. *Urban Studies*, 38: 231–237.

Imrie, R. 2012. Auto-disabilities: the case of shared space environments. *Environment and Planning A*, 44: 2260–2277.

Kitchin, R. 1998. 'Out of place', 'knowing one's place': space, power and the exclusion of disabled people. *Disability & Society*, 13: 343–356.

Lury, C. and Wakeford, N. (Eds) 2012. *Inventive methods: the happening of the social.* Abingdon: Routledge.

Macpherson, H.M. 2009. Articulating blind touch: thinking through the feet. *Senses and Society*, 4: 179–193.

Macpherson, H.M. 2010. Non-representational approaches to body-landscape relations. *Geography Compass*, 4: 1–13.

McCarthy, J. and Wright, P. 2004. *Technology as experience.* Cambridge, MA: MIT Press.

Norman, D.A. and Draper, S. (Eds) 1986. *User centered system design: new perspectives on human-computer interaction.* Hillsdale, NJ: Lawrence Erlbaum Associates.

Parr, H. 2008. *Mental health and social space: towards inclusionary geographies.* Oxford: Wiley-Blackwell.

Rose, G. 1993. *Feminism and geography.* Cambridge: Polity Press.

Spinney, J., Aldred, R. and Brown, K. 2015. Geographies of citizenship and everyday (im) mobility. *Geoforum*, 64: 325–332.

Suchman, L. 1987. *Plans and situated actions.* Cambridge: Cambridge University Press.

Suchman, L. 2011. Anthropological relocations and the limits of design. *Annual Review of Anthropology*, 40: 1.

Tolia-Kelly, D.P. 2006. Mobility/stability: British Asian cultures of landscape and Englishness. *Environment and Planning A*, 38: 341–358.

Urry, J. 2002. *Sociology beyond societies: mobilities for the twenty-first century*, Abingdon: Routledge.

Wallace, J., Wright, P., McCarthy, J., Green, D., Thomas, J. and Olivier, P. 2013. A design-led inquiry into personhood in dementia. In *Proceedings of ACM CHI'13*, conference, May 2013, Paris, France. New York: ACM Press, pp. 2617–2626.

Worth, N. 2009. Making use of audio diaries in research with young people: Examining narrative, participation and audience. *Sociological Research Online*, 14: 9.

Worth, N. 2013. Visual impairment in the city: Young people's social strategies for independent mobility. *Urban Studies*, 50: 574–586.

Wright, P. and McCarthy, J. 2008. Empathy and experience in HCI. In CHI '08 *Proceedings of the twenty-sixth annual SIGCHI conference on human factors in computing systems* (Florence, Italy, 5–10 April 2008). CHI '08. New York: ACM, pp. 637–646.

Wright, P. and McCarthy, J. 2010. *Experience-centred design: designers, users, and communities in dialogue.* New York: Morgan Claypool.

15 Feeling the commute

Affect, emotion and communities in motion

Emily Falconer

Introduction

Operating an understanding of mobilities that moves beyond the representational, this chapter will argue that how bodies relate beyond verbal communication is crucial to design, connections and community. The discussion will begin with an exploration of embodied affects in the social sciences more broadly, highlighting the potential for those who design spaces of mobility (transport, stations) to borrow established and emerging work within sociology and human geography, and setting the foundations for how a close reading of affect and emotion can shape our understanding of the design/mobility intersection. The chapter will then introduce the project on which this case study is based, outlining the methodologies employed to capture embodied affects 'on the move' during a train commute between Glossop and Manchester, UK.

This research was conducted in 2012–13 in the English town of Glossop, Derbyshire, UK, exploring notions of affect, affordance and interconnections as part of a project within the UK's Arts and Humanities Research Council's (AHRC) Connected Communities programme: "Revisiting the mid-point of British Communities: a study of affect, affordance and connectivity in Glossop".[1] The project explored how an affectual analysis of place, space and mobility can reveal a deeper understanding of how non-familial residents of Glossop connect and disconnect with each other. Highlighting contemporary residential migration patterns, practices of commuting and everyday mobilities, this focus asks how people's senses and feelings of community are constituted in relation to these mobilities and the affordances of particular spaces. This chapter focuses specifically on ethnographic observations of Glossop train station and the train carriage to shed light on the journey between Glossop and Manchester Piccadilly.

The chapter explores atmospheres of places and journeys, highlighting how they often involve multi-sensory mingling of soundscapes, and land- and city-scapes. Instances of sensory overload, pleasurable sensations, senses of calm, and the unfamiliar are presented, drawing on mobile interviews and participation observation. Illustrative examples presented include journeying with commuters as they traverse through the diverse multisensory atmospheres of an early morning train station, a familiar yet silent train carriage and a noisy, neon-lit, smell-ridden and body-filled metropolitan station. The chapter builds on earlier works of Bissell (2009, 2010; Bissell and Fuller 2011) to reveal the significance of mobility and the affordances of space in creating feelings and emotion of connection and separation, belonging and exclusion, community and individualisation.

The findings of this chapter reveal that amalgamating a study of affect and atmospheres within social and cultural contexts emphasises the role of design in informing the character of mobility, and ultimately contribute to community (dis) connection. The concluding section will invite further discussion into the very crucial challenge that not all bodies may experience affects equally, and that borrowing theories of affect and inequalities from across the wider social sciences can enable designers to shape mobilities based on enhanced understandings of social factors. There is great potential for a more nuanced understanding of affect in the social sciences to inform design and move beyond disciplinary boundaries.

Embodied affects: the potential for designers

Research into embodiment and emotions has benefited from the significant emergence of 'affect' in human geography. The 'affective turn' in the social sciences moves beyond previously constructed embodied approaches, as well as building on the politics of emotion within the social sciences, offering new insights into the atmospheric dimensions of human geographies (Thrift 2004; Anderson 2009; Bissell 2010). Affect expands the notion of the social to incorporate not just people, but a relational understanding to bodies, places, spaces, objects, lights, sounds and atmospheres, and advances our 'understanding of how embodied emotions and affect are intricately connected to specific sites, contexts and practices' (Jayne *et al.* 2010, 540). Affect is therefore conceptualised as a different way of thinking about emotions that occur outside of the body, where bodies are susceptible to a variety of external factors at a semi-conscious level (Thrift 2004). Recent accounts argue that the boundaries between emotions, embodied responses, sensualities, rhythms and flows become blurred as humans anticipate particular affective atmospheres and experiences (Edensor 2012). Literature that draws on the enhanced understanding of affect does not necessarily distinguish between the pre-cognitive affectual state and emotions as a consciously recognised sociocultural form (Edensor 2012).

The affective dimensions of ordinary life (Stewart 2007) highlight the capacity to affect and be affected by everyday momentary encounters, but also how these ordinary affects connect with the wider social and political world. I return to this imperative in my concluding discussion, where I argue that a sociopolitical understanding of affect could be potentially crucial to designers of space, bodies and movement. I argue that the concept of affect enables us to think differently about 'senses' of community, mobility and connection, and how this in turn influences how we may (re)design such spaces in order to facilitate feelings of belonging and community cohesion. There is significant scope for exploration of affect and emotion in designing mobilities, yet to date the links between these areas of study have remained limited.

Affecting mobilities: the study of the train journey

This is not to say there have not been in-depth studies of affect and commuting; indeed Bissell (2009, 2010; Bissell and Fuller 2011) has provided significant insight into the affective dimension of the passenger commute through the train journey. Bissell coins the term 'affective communication' (2010) when referring to the semiconscious communication between (unknown) train passengers during their journey. This communication is not verbal, and doesn't even have to include eye contact, but is significantly felt between bodies and the affective realm. For example, if a train unexpectedly slows down without warning, or grinds to a halt, the affective feeling of acquisition, dread, uncertainty or frustration circulates round the bodies in the carriage. Bissell argues that these affects are infectious, and the most significant communication between strangers moves beyond discursive registers: "(t)he precognitive, prediscursive affective registers of communication whilst travelling on public transport can significantly impact on the journey experience and what passengers can do" (Bissell 2010, 271). Affective registers circulate so we can feel the embodied emotions of our fellow passengers whilst on the move, and become attuned to their tensions, irritations, relaxation, and excitement. These affects are deeply related to the time of day, the season; for example the affects felt on a Monday morning commute may be experienced very differently to the uplifted mood Bissell describes of a Friday night train where some passengers are on their way to leisurely and social gatherings. Affective atmospheres within a train carriage are instrumental in both facilitating what people do, for example the practice of working on their laptops, drinking with friends, disrupting the relative peace of the carriage, but are also informed by these practices. Our prior knowledge and social positioning touches upon our sense of what is deviant, when noise is routine or disruptive. What can these shared, affective registers mean for designing mobilities in light of studies into community (disconnection)? How does design as a social relation translate into how we feel mobilities, the daily commute, the sensual variables of the train journey and station? Perhaps more importantly how can we situate these feelings in wider social and cultural contexts of change, community, connection and disconnection? There is great potential for a nuanced understanding of affect to illuminate the role of design in its production. The following section will provide a preliminary insight into how affects of the commute can be elicited.

Methodologies: eliciting 'affect, affordance, connectivity'

To understand the embodied and emotional experiences of community within a wider affective framework requires a mixed-methods qualitative approach. A mixed-methodological framework specifically aims to 'get at' making sense of affective experiences. Qualitative mixed-methods triangulation has been suggested to be particularly suitable for interdisciplinary research (Decrop 1999). Encapsulating affective experiences is indeed one of these 'complex purposes'. Geographical research has focused on the tactful 'choreography' of the mobile collective in public spaces, where bodies weave through streets, stations and markets and urban environments in close proximity (Edensor 2000; Bissell 2010). If we are to study mobility and affect, for example to apprehend how community residents feel the commute, we as researchers must also move through these spaces ourselves. The data for this project derived from movement, sensual observations, and banal commentary of changing surroundings, reflections and momentary stillness and pausing, as opposed to 'static', retrospective interview narrative. Three key methods were employed: mobile interviews, commuting interviews and creative methods.

Mobile interviews: walking, driving and mapping methods

There has been a strong emergence of methodologies in the human sciences that focus on the benefits of 'walking and talking' (Stals *et al.* 2014). The walking interview, where participants are encouraged to reflect as they move through particular spaces and environments, has been used to demonstrate the profound relationship between what people say and where they say it (Evans and Jones 2011). In order to capture how residents of Glossop felt about their environment, it was necessary for their reflections and observations to take place as we moved through both the town of Glossop and its surroundings. The mobile interviews took place in two stages. First, we met at the home or workplace of the participants, and conducted a brief, static interview, collecting general information about their lives, histories and everyday movements. We then asked participants to take us to a place which 'best represents community', or if this did not apply to them, a 'place which they felt was important to their everyday lives'. We then accompanied participants to their selected location, either on foot or in the car (the researchers drove, with participants in the passenger seat). During the mobile interview, data was collected in two ways: first, all narrative and discussions were digitally recorded; second, journeys were mapped via GIS tracking technologies. Mapping the journeys was of crucial importance; this allowed the research team to track the distance of participants' journeys in relation to their place of residence/work, as well as map out overlaps and intersections between participant mobilities, highlighting the most frequently visited places or routes. Mapping methods therefore became an integral part of data analysis, in order to identify where participants connected (or not). During the interview, photographs were taken by the researcher of places, buildings or objects visited or talked about.

Commuting interviews and participant observation

With a similar rationale to the mobile interviews, commuting interviews involved a specific journey to the participant's place of work. Glossop has become a settlement with a significant rise in workers who commute outside of the town to places of work in surrounding cities of Manchester, Sheffield and Leeds. During the commuting interview, which almost always involved a train journey to one such city, a member of the research team met with participants at the train station early in the morning and accompanied the participants on their journey, terminating the interview when they had reached their destination. During the interview, participants were asked to reflect on their journey and surroundings, and all narratives were recorded. In addition, the researchers took extensive notes, recording their own observations of the station (both Glossop train station and the train station of the final destination), the weather and the sensual affects (sound, temperature, feeling, moods, interactions between passengers). This was markedly determined by temporality, and the time of day becomes of crucial importance to observations of the commute. Researchers specifically chose to travel at different times of the day in order to achieve a more comprehensive understanding of how the rhythm of the commute has a direct impact on affective feelings of (dis)connection.

Creative methods: interactive play, art and visual activities

In partnership with a local community arts organisation, researchers on the project adopted a variety of creative methods in order to engage members of the community. This involved working in schools, care and residential homes, community groups and the main outdoor shopping market on a Saturday with local shoppers and passers-by. Participants were encouraged to express their feelings about community and the town of Glossop through drawings, photographs, mind maps, scribbles and doodles and creative writing. In addition, the researchers and community arts workers devised interactive board games ('Glossopoly'—a reinvention of the traditional Monopoly board game) which featured local places of interest (the station, the school, the town hall, the market, the butcher, the rugby club) to land on once the dice was thrown, and question cards to invite commentary and reflection of their experiences of such places.

The creative and kinetic methodology employed on the project proved integral to eliciting affect and how it relates to community and (dis)connection. The following extract from a walking interview exemplifies how reflection occurs within certain environments, evoked by particular affordances. Julia, a lone mother in her early forties, and I were in the station, at the very bench on which she reflects on her memory, as this recording occurred.

> We walked up the, up to the train station, my father and I, and we were looking at the trains, and he said 'Oh there's lots of trains from Manchester, every 25 minutes, that's very good', I said 'yes, I'd probably come up again on a Sunday, because it was a nice place'. And he sat me down on the bench on the, on the platform, and he said, um, well I've spoken to your sisters and they've all, all agreed that I'm going to re-mortgage the house and give you your share that you would get when we die now, so we can give you a percentage to put towards getting a mortgage yourself now, so my dad basically put a quarter of the money up so I could get this little terrace for (my son) and I, so we came here.

Here, Julia reflects on a memory of a conversation with her father, directly relating this conversation to a pivotal moment in her life trajectory, where important decisions took place, enabling her arrival into Glossop. What is crucial to this example is that this reflection took place on the bench highlighted in the story, at the station. For Julia, the station has become meaningful beyond the practical connections of commuting and travel. Thrift (2004) speaks of 'still points'—sitting down stimulates you into reflection in a mobile world. As we walk to the station, and sit on the bench this story gets remembered and narrated. Such rich data occurred in moments of movement and mobility, on our way somewhere, but also as we paused for moments of stillness and reflection during mobile interviews. I would suggest that this could not be captured in an orthodox interview narrative. Both the participants and researchers had to feel it, see it and experience it. This is, of course, dry and wind-free weather allowing. Acknowledging the weather is an increasingly important consideration in sensual ethnographies of social life and daily interaction (Mason 2016); and has until recently rarely been signposted as a significant factor in qualitative data collection. Had we chosen a rainy or windy day for the interview, this reflection may have been very different. It is these methodological insights which shape the case for a closer, affective exploration of commuting environments and the embodied emotions they facilitate.

Feeling the commute: affective environments of the train carriage

This chapter draws upon affective data from observations during the train commute from Glossop to Manchester, on a December morning in 2012. The following journeys took place over one particular morning, but at various times. This section explores the affective environment of the train carriage, as well as temporalities and rhythm, in order to highlight the role of design affordances in how the commute is 'felt'. Consider the following extract, illustrating the early morning atmosphere of the train carriage:

> Colin has been commuting for 17 years. He gets the 7.08 every morning to Piccadilly, changes to his workplace in Bolton. I meet Colin inside the carriage, and we sit opposite each other in an empty berth, speaking softly as not to disturb the incredibly quiet carriage and wrap our coats around ourselves. Everyone still has their hats, gloves and coats on. Colin points out that people sit on the same seat on the train every day. Same people, every day, same spot. People are quite possessive about it, and need their own space. But he doesn't speak to any of them. By and large they keep themselves to themselves. Where Colin sits, the same guy sits two seats away. Everyday. On the 7.08. He doesn't know him. The carriage is silent, separate, and people stay in their routinised, private spaces.
> (Field notes: Commuter interviews and participant observation, 2012)

The quiet stillness of the early morning carriage shapes bodies into silence and separation. Each has their own seat, familiar and comforting. We are cold, and do not remove our outer garments, very much contrasting the intended warm and welcoming design of other public spaces of hospitality (e.g. cafes, lounges, restaurants). Despite routines of familiarly (same seats, same people) these arrangements of people produce very little social interaction. Any communication occurs through the solitary medium of mobile phone technologies and social networking through Twitter. Whilst online communities and social media networks have a profound influence on the way we (dis)connect in changing ways (Kraut *et al.* 2012), the bodies present in the carriage are attuned to environmental affects and social norms which 'shrink' the body into individualised, contained spaces; the quiet, the dark, the cold. The choreography of affective bodies responds to the atmosphere of the carriage, producing an absence of connection.

Mobility, temporality and rhythm

Edensor (2012) argues that everyday temporalities circulate around our habits and bodily rhythms (mealtimes, tiredness), as well as fixed timetables, schedules, the conventional order of events and norms. We simultaneously desire rhythm regularity as well as difference, in our collective life. Elsewhere, I have argued that rhythm and temporality are crucial to how we embody experiences of mobility, tourism and long-term travel (Falconer 2013), especially with regard to eating and taste after prolonged periods of time 'on the road' and in transit.

Here I take this further to think how designers could work with this complex and multifaceted embodiment of rhythm, to see what bodies do when we are awakened in the same place each day. This is essentially very much embedded in both seasonal and daily temporalities which can change the affect of the commuting journey within a matter of hours. For example the commuters of Glossop are enveloped in silence during the early mornings of the winter months, and 'wake up' to work life when they arrive into Manchester Piccadilly train station, and light, dark, sound and affective atmosphere is crucial to this accidental design. This is not exclusively particular to these places, but can be said for multiple commuter-type stations and large, city intersections, yet how these spaces shape our embodied rhythms has yet to be fully explored:

> 7.08 am service: It's a dark, freezing morning. The station is full, but almost silent. A quiet, sleepy crowd. The platform is packed, but the quiet is very noticeable—voices are low, there are no announcements, no bright lights, no whirring of coffee machine. A Christmas tree glows in the corner.

> 9.08 am service: I return to Glossop station in full light to collect George to ride on the 9.08 service. The station is now in full daylight, and this appears to effect the sound—it is no longer the enveloped, slice of the dark that was the 7.08 service. People are talking, the automatic doors of the co-op (local supermarket) are opening and closing, background music is coming from somewhere. People interact on the platform, they are now travelling with each other, family members are on their way to Manchester for Christmas shopping perhaps, and there is a distinct lack of individual separatism that

there was a few hours earlier. Garry and I chat on the platform for some time about the project, and my work generally. We don't need to lower our voices, self-consciously as not to disturb others, like I did with Colin.

We discuss the sensory differences of Piccadilly station. Arrival, security guards everywhere checking tickets. You emerge and it's a world away from Glossop. There is the sensory experience of moving media, bright neon lights, advertising, endless hollow announcements, untrusting ticket checks—you are in a city!

(Field notes: Commuter interviews and
participant observation, 2012)

This reading evokes the question of how time and temporality (including seasons, annual festivals, temperature and weather) come together to affect how we experience the commute. Emerging from the rural sleepiness of a cold, dark morning into the stimulus of an inner city major transport hub has profoundly affective impacts. Anderson (2009) asks how atmospheres 'envelop' and 'press' upon life. As we arrive into Manchester Piccadilly we are ejected from the thick atmosphere of silence and familiar faces and routine into a stimulating crowd, awakening all senses. Mindful of the multiple sensual variants that create the 'affective envelope', designers can facilitate and simulate affects of being closed, cocooned in thickness of darkness and silence, or lost in a sea of stimulus.

Moreover, these temporal sensations impress upon moments of (dis)connections, and have further implications for thinking about communities and how they may be understood:

We look around the carriage, I ask if this is normal. George explains this is quite quiet. People tend to meet and sit together and have a coffee. There is a difference which train you get.

George explains that the 8.03am is a completely different atmosphere, and reminiscent of when he used to live in London. Crowded, sleepy, fixed on themselves, self absorbed, they think about what is going on in Manchester, not Glossop. The atmosphere is more competitive, more separate. The later service is far more relaxed. I look around. It feels it. But George says this is not place specific, the later services feel more quiet and friendly even on the tube in London!

(Field notes: Commuter interviews and
participant observation, 2012)

Glossop and Manchester appear to be two distinct worlds, two distinct relationships. You can see people transitioning from one to another. However, as exemplified, the time of day greatly influences the affects produced, and determines whether connections are made, and with whom. It can therefore be argued that it is not only spaces, but temporalities, which smooth connections and interactions, or conversely encourage introversion, during the commute: allowing eye contact, talking loudly, 'warming' up to our surroundings or remaining contained within ourselves. This chapter argues that designers of transport, stations and leisure spaces can be greatly influenced by the seasonal and daily rhythms of everyday life and affective atmospheres.

Concluding discussion: taking further steps

The disciplines of human geography and design have much to learn from each other. Combining the study of affect and emotion in the social sciences and human geography with studies of community, mobility and connection can provide valuable insights into the realm of design. Thrift, in a powerful discussion "towards a spatial politics of affect" (2004), acknowledges the alliances between the social sciences and the arts; the "engineering", he claims, produces both theoretical and practical knowledges "which can simultaneously change our engagements with the world" (2004, 75). This chapter illustrates how affective environments, rhythm and temporality are deeply intertwined with the experience of the train commute, with specific focus on whether these affects enable or inhibit connections and interactions. Embodiment, emotion, affect and affordance can tell us about communities, and how community studies are evolving, but have further implications for those who design such spaces of movement and belonging. If seasonal and daily rhythms have a significant impact on whether people connect or disconnect—is there scope for designers to work with these affects? Affect is transmitted from body to body—we feel the commute collectively. Acknowledging these affects can determine whether stations, carriages and waiting rooms can be kept calm, light, dark, warm to fit in with our bodily rhythm (for example a dimly lit carriage when the season is dark may keep it enveloped and sleepy, 'waking up' in the spring).

However, whilst this chapter calls for a communication between the disciplinary boundaries of design and social science, a simple recommendation of how to work with these affects (soft sounds, chairs facing inwards, warm platforms) implies that all bodies experience affects in the same way. This is not the case. Affect is, by its very slippery nature, ambiguous. Anderson insists we need to embrace these ambiguities, not holding on to any certainty, concrete conclusions or "exaggerated trust" (2009, 78). This is problematic for those designers who might seek to shape the character of mobility primarily because the affects produced through the arrangements of bodies and things depends so much on how the bodies relate to them. What has not been explored in this chapter, but what remains crucial for further analysis into amalgamating affect and design, is the extent to which affect is experienced equally with regard to gender, race, class and other social inequalities.

Elsewhere I have argued that sensual affects of dirt, cleanliness and disgust can deeply affect the inclusion (or otherwise) of certain social groups, specifically with regard to gender and social class (Taylor and Falconer 2015). Similarly, there has been recent work into how affects of fear of public transport are deeply gendered (Loukaitou-Sideris 2014; Hewitt 2014). This line of thought follows Ahmed (2000, 2004), who claims that even unconsciously experienced affects, which cannot be recognised or attributed to a direct emotion through cognitive understanding, can be evoked by past encounters embedded in social and cultural histories, and inform the narratives of future embodied and emotional encounters. The process of designing mobilities and affect does not occur in a social vacuum—some bodies will be excluded and included, unequally, and this is a key concern for designers who need to incorporate an interdisciplinary approach to design by borrowing theories of affect and how this relates to social and cultural politics. To be clear, this responsibility is to work closely with sociological and cultural geography theories relating to unequal access to spaces, in order to be wary not to reproduce exclusive spaces and conflicts through producing particular affects. Be that as it may, I argue that designers can, and indeed should, work in conjunction with the emerging theories of affect theorised by social scientists and political geographers in order to consider the importance of bodily relations to inclusive spaces, (dis)connection and notions of community. I suggest that understanding how bodies relate to artefacts beyond verbal communication is crucial to design, connections and community.

Note

1 The *Connected Communities* Research Programme aims to "understand the changing nature of communities, in their historical and cultural contexts, and the value of communities in sustaining and enhancing our quality of life" (www.ahrc.ac.uk). The town of Glossop was seen in the mid-twentieth century as being a place of an autonomous community lying between the rural and the urban but is now widely viewed as a commuter settlement with many residents spending much of their daily lives outside of the settlement in a series of urban and peri-urban spaces. A large number of Glossop's population now commute to spaces of leisure and employment in surrounding cities of Manchester, Leeds and Sheffield. With thanks and acknowledgements to Principal Investigator Professor Martin Phillips, University of Leicester.

References

Ahmed, S., 2000. *Strange encounters: embodied others in post-coloniality.* London: Routledge.

Ahmed, S., 2004. Collective feelings or, the impressions left by others. *Theory, Culture & Society*, 21: 25–42.

Anderson, B., 2009. Affective atmospheres. *Emotion, Space and Society*, 2: 77–81.

Bissell, D., 2009. Conceptualising differently-mobile passengers: geographies of everyday encumbrance in the railway station. *Social & Cultural Geography*, 10: 173–195.

Bissell, D., 2010. Passenger mobilities: affective atmospheres and the sociality of public transport. *Environment and Planning D: Society and Space*, 28: 270–289.

Bissell, D. and Fuller, G., 2011. *Stillness in a mobile world.* London: Routledge.

Decrop, A., 1999. Triangulation in qualitative tourism research. *Tourism Management*, 20: 157–161.

Edensor, T., 2000. Moving through the city. In *City visions.* Ed. D. Bell and A. Haddour. London: Prentice Hall, pp. 121–140.

Edensor, T., Ed., 2012. *Geographies of rhythm: nature, place, mobilities and bodies.* Farnham: Ashgate Publishing.

Evans, J. and Jones, P., 2011. The walking interview: methodology, mobility and place. *Applied Geography*, 31: 849–858.

Falconer, E., 2013. Transformations of the backpacking food tourist: Emotions and conflicts. *Tourist Studies*, 13: 21–35.

Hewitt, T., 2014. Beyond bright lights and security cameras: Re-gendering Melbourne's public transport system. *Planning News*, 40: 24.

Jayne, M., Valentine, G. and Holloway, S.L., 2010. Emotional, embodied and affective geographies of alcohol, drinking and drunkenness. *Transactions of the Institute of British Geographers*, 35: 540–554.

Kraut, R.E., Resnick, P., Kiesler, S., Burke, M., Chen, Y., Kittur, N., Konstan, J., Ren, Y. and Riedl, J., 2012. *Building successful online communities: Evidence-based social design*. Cambridge, MA: MIT Press.

Loukaitou-Sideris, A., 2014. Fear and safety in transit environments from the women's perspective. *Security Journal*, 27: 242–256.

Mason, J. 2016. Living the weather: project summary. Available at http://www.social sciences.manchester.ac.uk/morgan-centre/research/research-themes/relationalities-friendship-and-belonging/living-the-weather/, accessed 16 June 2016.

Stals, S., Smyth, M. and Ijsselsteijn, W., 2014. Walking & talking: probing the urban lived experience. In *Proceedings of the 8th Nordic Conference on Human-Computer Interaction: Fun, Fast, Foundational*, October, pp. 737–774. New York: ACM Press.

Stewart, K., 2007. *Ordinary affects*. Durham, NC: Duke University Press.

Taylor, Y. and Falconer, E., 2015. 'Seedy bars and grotty pints': close encounters in queer leisure spaces. *Social & Cultural Geography*, 16: 43–57.

Thrift, N., 2004. Intensities of feeling: towards a spatial politics of affect. *Geografiska Annaler: Series B, Human Geography*, 86: 57–78.

16 Drawing mobile shared spaces
Brighton bench study

Lesley Murray and Susan Robertson

Introduction

Drawing plays many roles in relation to design, recording, reflecting, capturing and creating situations and conditions that are measurable as well as those that are perceived and conceived. Drawing is a 'frame of the imagination' (Farrelly 2011) allowing us to envisage relationships that are not usually visible and to consider and test both the probable and improbable. Architects and urban designers use drawing to think about spatial arrangements in the city and to develop the mental constructs of potential occupations in urban spaces. Drawing has been used as a method of data gathering and a pedagogical research tool (Bagnoli 2009), and also has been incorporated into the field of mobilities (Cresswell 2006; Sheller and Urry 2006; Urry 2007) with the emergence of mobilities design (Jensen 2014).

Taking inspiration from Appleyard, Lynch and Meyer's *The View from the Road* (1964), this chapter illustrates the potential of drawing to both 'capture' and to interrogate the complex relationship between the design and mobile practices of street space, particularly street environments that have been designed to re-imagine the relationships between walkers, cyclists, car users and others in mobile space. Our examination of a specific shared street space—New Road in Brighton—aims to contribute to new ways of seeing the interfaces of design and situated mobilities: movements and their meaning that are contextualised in social and cultural space. We seek to capture how the mobile practices performed within designed street spaces are the means by which built form is known, whether by the moving eye scanning space, by the feel of surface through the feet or via the body of the vehicle, the effort or ease in covering the 'ground', or the shifting proximity and arrangement between people and surfaces. In particular, we deploy material from a 24-hour 'bench survey' in New Road to investigate these ideas.

Understanding designed street space

Drawing upon Lefebvre (1991, 2004) we are interested in the ways in which urban encounters are situated within streetscapes that echo power hierarchies and are mediated by the contingencies of social space. Although we are concerned with materialities and their assemblages, we also recognise the importance of the differentiated subject and the significance of social difference in producing urban space. Lefebvre's (1991, 2004) conceptualisation of space allows attention to intersections between the impositions of differential imbrications of power on public space, with embodied experiences and urban imaginaries. In particular, 'rhythmanalysis' (Lefebvre 2004) allows a focus on embodied practices and a mindfulness of the relations of power in which they are situated. Mobile practices are embodied and multisensory, while at the same time urban spaces are produced through sensory experience (Degen and Rose 2012). Nuanced accounts of the situated and embodied nature of mobile practices in street space, we argue, are revealed through drawing. We can begin to unravel practices of power, mobility and space by attending to both sensory experiences and the ways in which movement is produced through it (Howes 2006). Degen and Rose (2012) argue that both academic and policy discourses around the use of urban space are based on the experience of design with less attention to the sensory experience of designed spaces.

Our interest in this chapter is in the intersection of embodied experience and situated urban design. We draw upon studies of the significance of sensorial encounters in the twenty-four-hour city (Adams *et al.* 2007), seeking to examine the city as experienced by all senses, not solely the visual. We consider how sensory experiences become mediated by "different and shifting spatial and temporal practices" (Degen and Rose 2012, 3), by practices of spatial mobility and memories of previous experiences of place. Sensorial experiences of space are produced through the encountering of the material and cultural characteristics of the space; and by the ways spaces are configured and felt through seeing, hearing, smelling touching and tasting. In turn, emotional responses to space are dependent on these sensory moments. The contention here, therefore, is that we need to develop an understanding of the sensory, and thereby emotional dimensionality of space through a transdisciplinary approach that draws from both the social sciences and arts and humanities through geographies of the senses (Degen and Rose 2012); sociologies of the senses (Simmel 1997 [1907]); and sensory design and architecture (Malnar and Vodvarka 2003; Pallasmaa 2005). Sensory ethnography is used as a framework for understanding multisensory spatial practices where "the senses are not separated at the point of perception, but culturally defined" (Pink 2009, 13).

Researching street space through drawing

In order to develop understandings of emotional and embodied encounters in street space, we argue that drawing offers an alternative way of seeing to more traditional means of 'capturing' spatial encounters. We are concerned here with 'visuality': an "understanding of images as meaningful objects central to symbolic and communicative activity that is core to many theorizations of contemporary visual culture" (Rose 2014, 32). Rose (2014) argues that researchers using visual research methods have paid little attention to visuality. In drawing through a continuum from fieldwork to analysis, we direct awareness to the symbiotic relationship between the visual method and the visual culture where, as Rose (2014) contends, there is potential for a shared understanding of images as communicational tools. Public and mobile spaces are often evaluated using traditional research methods, which overlook the nuanced use of space and the intersections between social and material interactions that influence its use in a particular way (DfT 2009; Gehl Architects 2010).

Our approach seeks to fuse innovative methods from both social science and arts and humanities to explore situated encounters in the street. The wider research on which this chapter is based adopted an interpretative and interactionist approach drawing from methodologies and methods in both the social sciences and architecture.[1] The research is based on a theoretical frame that incorporates everyday social, embodied and material experiences of public space; observations and readings of public space by users and non-users; and points of friction in the material/social interface (de Certeau 1984; Goffman 1966; Lefebvre 1991). Given that our research focus was upon interactions in relation to the inhabitation of space, we were as interested in the moving in and 'resting in' or 'being in', as the moving through.

Revealing rhythms through drawing

The multisensory practices of space are revealed in different ways at different scales of seeing. De Certeau (1984) illustrated the disparities of vision of the panoptic gaze and the 'zoomed in' street level gaze of the ethnographer. Lefebvre (1991) on the other hand asks us to zoom out from this micro scale, to the 'window', where an optimal gaze gives us a view of the micro-socialities of social space but also of the context in which these practices take place. Zooming in and out therefore illuminates the disjunction between the overall pattern of movements seen from a distance and the quirks of individual movements when seen close up/in detail—and the differences of space/time.

Jensen (2014) has illuminated the role of storytelling in mobilities design as a way of interpreting and producing the urban environment, from Vannini's (2012) narrative of theory and ethnographic findings to Marling's (2003) inspiration in the 'songline' as a means of navigation of the environment by Australian aboriginals. Storytelling is integral to ethnography, as it is integral to everyday life. It is partly for this reason that the two-dimensional image is considered "completely inadequate for capturing the dynamism of a mobile situation" (Jensen 2014, 28). Nevertheless, Jensen (2014, 28) acknowledges the role of the image as an "active design tool" using Cresswell's (2006) elaboration of the production of mobilities through representation and analysis of the work of photographers Marey and Muybridge. Other photographers, pioneers of ethnographic urban study, such as Jacob Riis (1890), have moved beyond the image as flat representation to its practice in mobilising concern for new urbanites living in poverty (Green 1985). Similarly, architects and other spatial designers, such as urban designers and landscape architects, use images as a design tool in this way, although more commonly in combination with other representational methods such as drawings.

Landscape designer and academic James Corner produces highly complex drawings from a composite of images that at least in part provide a sense of the experience of place at a specific moment. For example, burning the stubble in a field is captured through the inclusion of a plan of the contours of the land, a photographic image of burning material that evokes the smell, a drawn scale of temperature and the wind direction shown in relation to the plan (Corner and MacLean 1996). The drawings suggest at least one register by which measurements may be made (temperature or wind speed or direction etc.). A composite, multiscalar image is produced digitally. The engagement of making by hand is less removed and more open to subtle inflections and these may be seen as valuable, especially in examining relationships in detail and at the scale of urban spaces. We look to understand architectural space and form through the closer connection of body to form and space, in both the kinaesthetic and the imaginary senses. We are partly concerned here with Jensen's (2014) use of drawing (from Dovey 2010) as disentangling. Drawing is also, as Ingold (2011, 177) argues, "fundamental to being human—as fundamental as are walking and talking . . . because even without a pencil we are drawing with our bodies—making paths, gesturing, communicating, leaving trace or trails". From this, Jensen (2014) argues that diagrams help us to articulate thoughts, to produce thoughts and help form conceptual frameworks.

However, the diagram, used as a tool in spatial design, is a highly ambiguous and contested form of communication, with the potential to mean almost anything to almost anyone and even to become unfathomable to the author of the diagram.[2] The diagram may be representative but not 'accurate' or objective; diagrams may indicate an existing situation or a future, imagined situation—or even both simultaneously, thus revealing or suggesting shifts in both time and place. While Jensen (2014, 42) is more interested in contemplating the diagram as a design tool, "as practical tool and powerful mental technique", the element of experimentation is acknowledged as outside his normal way of working. As we have just mentioned, drawing is as natural as walking and talking and may be considered an equally essential communication tool. If "most contemporary architects love to draw but hate to write" (Ingold 2011, in Jensen 2014, 45), is this in part because of the relative openness of interpretation of drawings as opposed to the pressures of being clearly understood with the written word? The diagram, as opposed to a realistic representation of space, also allows more scope for the imagination. In Jensen's (2014, 42) discussion there are clearly acknowledged tensions between thinking of diagrams as "representational" or "vehicles for thinking"; as he acknowledges, the diagram can look deceptively simple and indeed simplistic, lacking "multisensate and emotive dimensions".

Diagrams can be used in multiple ways, frequently as visual clues and reminders. But there is a danger that the aesthetic of such diagrams may deceive and become generators of projects that have their own internal logic without taking account of real contexts. Jensen (2014, 45) refers to "design analysis" in his discussion about the use of his diagrams but this seems to be a contradiction in terms. The sequence of the spatial design process might be simplified as follows: observation, documentation (mapping, filming, photographing, etc.), analysis and design response, incorporating the manipulation and development of a range of media. An in-depth understanding of an existing situation and responding to it in terms of design are interdependent aspects of the process of design. Diagrams tend not to take account of the micro-mobilities that take place in negotiating spaces—that is between bodies, bodies and things both static and moving—so we must develop other images and drawings that can take account of these effects.

As we indicate below, our scale of vision and the methods we employ to capture it allow attention to the rhythmanalysed negotiation of space (Lefebvre 2004; Vergunst 2010). For Lefebvre (2004) rhythm originates in the body—in the rhythms of the body, the breath, the heartbeat. The body under capitalism is a central theme, but rhythmanalysis also allows a contextualisation in relations of power: that is, how embodied rhythms intersect with rhythms of authority and control. Lefebvre (2004) talked of the slowing down of rhythms at night, the normative rhythms of urban encounters. And so the street is ordered through particular authoritative controls even where "the orderly street has given way to a multitude of interweaving routes and improvisational lines" (Vergunst 2010, 381). Here we make visible a more consistent rhythm as time and space converge in a reconfigured differentiation of space in which the normative diurnal rhythms are disrupted. Our mode of capture is visual with an understanding of images as 'meaningful objects' that produce visual culture (Jensen 2014; Murray and Upstone 2014). Rather than flat and static representations, images become part of mobile practice. Images are implicated in the space and the mobile doing that are part of it. As Lefebvre (2004, 25) intended, rhythmanalysis brings together "diverse practices and very different types of knowledge" and thus in this chapter we bring together ethnographic and graphic knowledges to reveal the assemblage of urban encounters (Vergunst 2010).

Drawing rhythms

In *The Manhattan Transcripts,* Tschumi (1981) explores a tripartite representation of engagements with space, placing photographic image, drawn plan and diagrammatic representation of movements in the space alongside each other—allowing these three modes to be read more or less simultaneously and thus suggesting the over layering of different registers that combine to create a multidimensional understanding of place and time. This method, in its fragmentary, episodic but also sequential sensibility has a strong relationship to the cinematic experience of automobilities first explored in detail by Appleyard *et al.* (1964). In Lynch's earlier work (1960), diagrams describing 'mental maps' were constructed from questioning hundreds of people about how they navigated cities; and the diagrams thus formed an 'average' or 'typical' mental map. In later work, Lynch (1981) adds the representations of power relations and their spatial implications to his diagrams. He describes the changed pattern of cities from earlier diagrams radiating from one centre, to a more complex machine-like system of multicentred interconnected enclaves. Lynch (1981) indicates that connections may vary in terms of their significance, pace and extent. Such "diagrams of extension … represent the spacing and interval between objects, a matter of great importance in the modern city with its accelerated personal mobility" (Dulić and Aladžić 2016, 88).

While Lynch (1981) describes enclaves as mono-functional, it is possible to view specific street spaces as multifunctional: spacing and intervals are over layered in both time and space, allowing us to reconsider the rhythm of the flows in a particular place. Drawing allows us to present intersecting rhythms, to show the ways in which, just as embodied rhythms coalesce, abut and diverge so too do they intermingle with the rhythms of the material. At our case study bench on New Road (see Figure 16.1), we can consider, for example, the wet or dry bench; the duration of micro-indentations along its length from those who come into contact with the bench; rubbish moving in the wind, being picked up by different people and dropped by others; objects are left, forgotten, remembered and retrieved. The potential to understand and respond to the desires of occupants, at different scales, is an area of investigation that is, so far, underexplored.

In the meantime, the immobilising tendencies of architectural drawings are implicated in the gap between Lefebvre's representations of space and mobile practices (Robertson 2007). So we are interested in exploring methods that allow us, as designers and social researchers, to "unwrap the bundle" (Lefebvre 2004, 19). Lefebvre sets up the beginnings of a list of categories, or concepts, that may be examined in an analysis: "Repetition and difference; mechanical and organic; discovery and creation; cyclical and linear; continuous and discontinuous; quantitative and qualitative" (2004, 19). These can become the starting points for drawing a rhythmanalysis of New Road as an experiment, given that "the ways we capture and represent mobilities are reflections of how we comprehend and understand the phenomenon of mobility at a very profound level" (Jensen 2014, 27).

Researching designed street space through drawing: Brighton's New Road

In order to illustrate the potential of drawing to 'capture' and interrogate the complex relationship between the design and the mobile practices of street space we now turn explicitly to research focusing upon New Road in Brighton, United Kingdom. The 'shared space' scheme in New Road was designed to re-imagine relationships between walkers, cyclists, car users and others in mobile space. Benches are a key design feature of streets made for liveability and New Road is no exception. Consequently a key component of our methodology was a 24-hour 'bench study'.

Whilst Jensen (2014) discusses a range of representations of mobilities, these refer predominantly to singular modes of mobility—to singular rhythms. In contrast, the challenge in our research was to use drawing to explore the myriad and situated rhythms of the New Road street space. Although architectural drawings are generally static representations of buildings in space and have been critiqued for their limitations in allowing for movements and fluidity, we argue that the slow process of drawing can facilitate the analysis of mobile practices at a more

Figure 16.1 The bench in New Road.

measured pace in order to make emotional engagements visible. We seek to exploit the slowness, or slowed-downness, of 'shared space' to look at moments through drawing and at the same time attend to the detail of the context in which moment to moment micro-mobilities take place as a street narrative.

As we have indicated above, our reflections upon intersections of multiple mobilities draw upon Lefebvre's (1991, 2004) theories of social space to consider the representations of space, which may be emblematic of, for example, particular power relations; the spatial practices, which are the mundanely or routinely experienced aspects of space; and representational spaces, which are the imagined aspects of space that offer potential for appropriation. We consider the shared space of New Road in Brighton as material form and space (Merriman 2006) and as a site of experience and cultural image (Cosgrove and Daniels 1988). Space is regarded as a 'medium' not a 'container', such that space and action are inseparable (Tilley 1994).

Traditionally, transport planning has been concerned with maintaining networks of flow, and urban architecture has often stopped at the kerb. Similarly, studies of urban street spaces have tended to focus on their sustainability in terms of different modes of transport (DfT 2009). However, there has been less emphasis given to the potential of designed spaces in producing knowledge of their intricate social relations. The speediness of everyday life evident in most urban streets prohibits a close examination of these relationships. However, the approach of 'shared space', in which physical divisions between users are removed, allows for the slowing down and freezing of urban movement. Through the reconfiguration of the material space is a 're-staging' (cf. Jensen 2013) for the re-negotiation of mobilities. Hans Monderman, the designer most closely associated with the idea of shared space, proposed such a re-compositioning of street space in a way that gave responsibility to individual users to negotiate their use of the space with other users. In the case of New Road in Brighton, drivers, cyclists and others who move at different speeds through the space negotiate their speed, direction and dwelling in the street with those who spend more time there, whether sitting or playing, drinking, dancing or performing music. It is this intersection of the materiality of space and its social and cultural mobile practices that makes designed street spaces of particular interest to critical mobilities studies.

The bench central to our study is timber-clad and under-lit at night, is a relatively comfortable resting place and is, materially, without any of the defensive architectural features used to prevent homeless people using it for sleeping, which are becoming commonplace on many 'public' benches (Omidi 2014). Nevertheless, the bench is, of course, more than its materiality and is imbued with cultural meaning as well as governed according to its socio-political context. The New Road bench is itself the product of mobility. The street benches are an integral part of an overall design by Copenhagen-based Gehl Architects (2010) and the benches themselves were designed and manufactured by Lancashire-based company: Woodscape. It is likely that the hardwood and metal fixtures themselves travelled from further afield. These originating qualities of the bench can be traced and become part of its on-going narrative. As the hardwood weathers and transforms, it follows its own trajectory, whilst it becomes part of the narrative of those who sit, lie, walk, run, climb and skateboard on it.

The fluctuating significance of the materialities of the bench is also implicated in the spatial practices of the street. We watched the bench in New Road over a 24-hour period in June 2014. Our ethnographic study comprised a mix of design and mobile methods (Büscher *et al.* 2011, Fincham *et al.* 2010). We observed the bench and recorded activity, on and around it, at regular intervals using field notes, video, photography and drawings. We also carried out mobile interviews with people using the bench. Although the methodological approach here appears to give prominence to visual methods, this does not mean that we privilege this sense over others but rather we approached the project with the understanding that all senses are interconnected and that the visual reveals other sensory engagements with space (Pink 2009; Rose 2014).

Hence, our focus is first on *capturing* this multisensory experience through drawing. Figure 16.2 presents a selection of sketches made during the 24-hour study. The drawings of people sitting, talking, drinking and moving are tracings of the mobile practices around the bench, which have an ephemeral quality. They are textured in a way that implies movement and interaction. Without seeing faces we can gauge mood through body shape and posture. This meticulous yet undetailed stage of the process is the first step of any design investigation, paying very careful attention to all the occupants, human and non-human, of the space under consideration.

If we look at some of the drawings made over the 24-hour bench study (Figure 16.3) we can identify some of the rhythms Lefebvre (2004) discusses. The

Figure 16.2 Sketch of New Road bench during 24-hour study.

'measure' of the constituents of rhythms may be identified in drawings, such as the time that the plans and sketches were made—sometimes shown, sometimes not.

The base drawing shows the benches, the trees, the colonnade and other 'static' elements. But of these elements, even the very surface of the colonnade, for example, will change over time: accumulating very fine layers of dirt; becoming damaged in the form of scratches and chips; appearing to change shape and tone as the light conditions change and so on. We can identify 'polyrhythmia' where the concurrent and simultaneous rhythms of the bench are linked to its materiality and meanings along with the 'natural' rhythms of each occupant whose bodies may be considered as terms of reference: the bench is something to sit or sleep on. It is also a refuge and 'a dirty old thing that attracts noise and trouble'. The dominant or 'staging' rhythms also change: for example, a police presence to check on the space's homeless people pervades the space in the morning. In the evening, the usual dominant modal rhythms of the street become less relevant as the rhythms of automobility are dampened by the volume of people during the day and are in direct conflict with pedestrians at night when young people appropriate the space. We also identified 'arrythmia' ('abnormal' rhythms), for example, a man making and selling origami and an older couple permeating the festivities at night. These practices stand in contrast to more pervasive rhythms of the street.

Drawings were created to encourage a 'drawing in' by asking the eyes to rove over, demanding a reading and a decoding through imaginative interpretation and a 'drawing together' of relationships between representations that suggest the particular kinds of reality that are then constructed. Composite images were

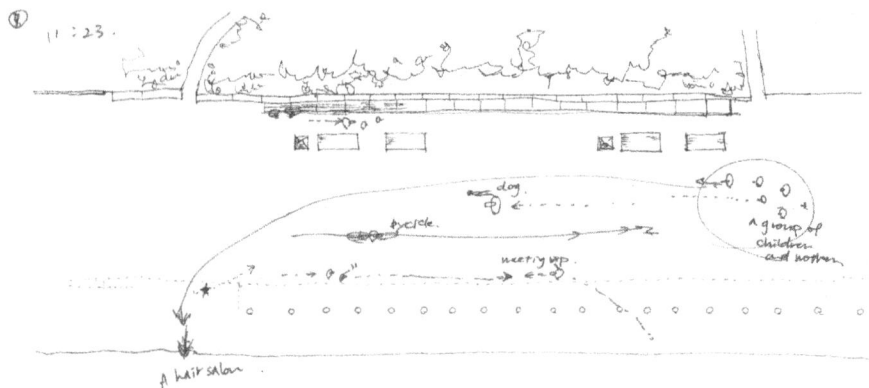

Figure 16.3 Plan drawings of movements in New Road.

made, overlaying drawings with other images in order to illuminate visible and invisible measures. For example, the constant grid of Lefebvre's (1991) 'spaces of representation' can be provided by an Ordnance Survey map, with all the connotations of place and time according to geographical longitude and latitude that are universally imposed. Other elements illuminate relationships in the shared space, suggesting something more poetic than the straightforward and consistently plan views that (modernist) planners work with. We begin to appreciate the complexity of the space and the myriad embodied and disembodied rhythms within it. We think of the bench as the provocation allowing the space to be variously occupied and situated as closely as possible for minimal additional intervention—a haven in an otherwise hostile milieu.

The temporal rhythms we seek to explore have different ranges. At one level we have worked with cartographic juxtapositions of older Ordinance Survey maps (e.g. the 1974 OS map which predates the New Road redesign) and a relatively up-to-date digital map (Digimap), in order to reveal the changing structure of the street through time. We have also sought to capture how daily rhythms of the street and the bench are configured in different ways: Figure 16.4 depicts a drawing which developed from a selective set of conditions in relation to time and place. The weather and time are described through the depth, length and direction of shadows shown at intervals. Traces of occupations are shown indicating mobilities alongside the relative speeds of bicycles and cars passing through; these are shown at the scale of the street and also at the micro-scale of one part of the bench.

From here, following Appleyard *et al.* (1964), and in appreciating the temporal element of the space, we can develop more abstract representations of the interactions of space, time and mobilities. Returning to the fieldwork data, the observational notes and interviews carried out over the 24-hour study, we began to think about how the intersecting rhythms of the street could be represented according to Lefebvre's triptych of space. We took inspiration from Alison

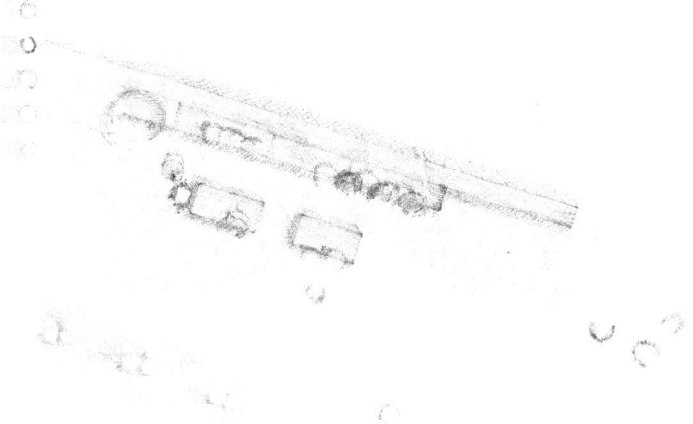

Figure 16.4 Drawing the space and time of the bench.

Turnbull's (2013) drawings, which are drawn directly onto graph paper. We began with a 'graph' plotting the rhythms of the bench over the 24-hour period. Vertical graph lines represented linear time in hourly intervals, starting at the beginning of the survey period (07:00) on the left-hand side of the page. Horizontal undulating lines represent volumes of street users; generation, gender, passing through; automobility; and sitting. They illustrate the ways in which the street is practised as well as the mobile norms that produce these practices, for example gendered norms produced through the gendering of mobile spaces (see Priya Uteng and Cresswell 2008).

So we used both paper and traces made on it to represent the tripartite rhythms of the street. For representations of space, the dominant authoritative rhythms of the street are imagined in grid lines. Spatial practices are revealed in the linearity of pencilled lines and representational spaces in the potential of irregular pencil marks. The drawing suggests specific relationships between sets of rhythms and reveals how manipulation of any of the spatial and temporal elements captured will affect other elements. The closer we get to the detail of the rhythms of these mobilities, the clearer the choreography of interactions becomes, and the more open it is to analysis.

A final stage of work consisted of a drawing that combined both the hand-drawn rhythms of the bench and the graphic plotting of temporal rhythms through the 24-hour research period. Figure 16.5 is both a 'map' and a 'day in the life of' the street bench. But this is only the beginning of the interrogation of the space. The drawings provide new 'ways of seeing' that bring together selected assemblages of actors and mobilities that are open to interpretation as images. We see the ways in which different rhythms are not only present in the space but are the space.

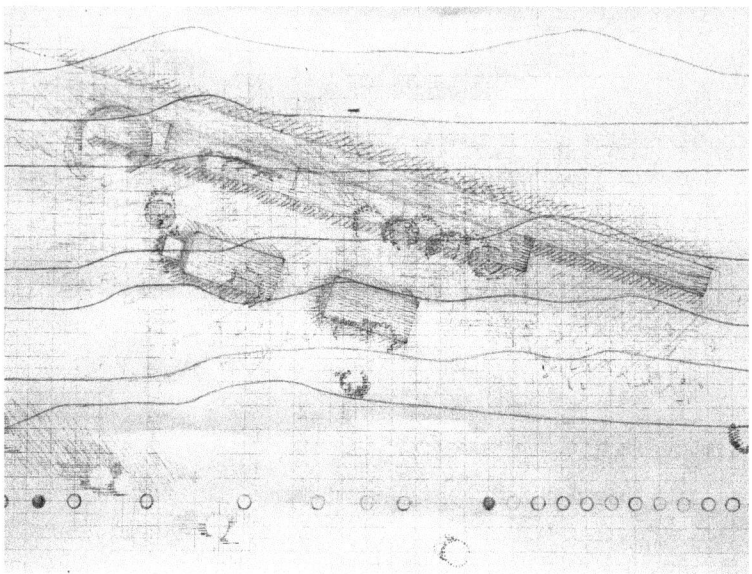

Figure 16.5 Intersecting spatial and temporal rhythms.

Conclusion

Drawings are speculative in ways that are aimed to provoke thought about how spaces have been made and how designed elements may operate in practice. Drawings reveal aspects of the bench study space that may otherwise go unnoticed and thereby can provide a discrete contribution to the emerging field of mobilities design. In particular a focus on drawing foregrounds an aspect of design that is insufficiently incorporated within other aspects of architectural and urban design, although architects such as Tschumi (1981) have discussed these ideas previously and have even developed designs in response. Appleyard *et al.*'s (1964) work in drawing the road also offers much in relation to the experience of mobile space and yet is undervalued.

Further, however, one of the more revelatory aspects of drawing is that it offers potential to invigorate or reinvigorate ethnographic research. Drawing was one aspect of our transdisciplinary study of a mobile space, which set out to uncover aspects of this space and specifically the negotiation of urban encounters. Scrutinised in relation to alternative ways that were used to capture the space over the 24-hour period, drawings become part of a mosaic that offers a rich and insightful overview of the space, from varying angles. It is through bringing together research tools that we can conceptualise interactions between the political, material and embodied aspects of the space. In recognising the creative role of measuring and drawing in design, thinking about the production and construction of mobile spaces may lead to more intentionally created urban landscapes of mobilities.

Acknowledgements

This research was funded by the School of Applied Social Science Research Development Fund, University of Brighton. We would like to acknowledge the students on the University of Brighton Masters in Architectural and Urban Design who contributed to this chapter and especially to Blake Lewis for Figure 16.1 and Mami Masuda for Figures 16.2, 16.3 and 16.4.

Notes

1 This chapter is based on an ongoing study of New Road in Brighton undertaken by the University of Brighton Masters in Architecture and Urban Design and through a 24-hour ethnography of the bench that runs along New Road.
2 See Garcia (2010, 22) for a list of possible interpretations of the diagram in the fields of spatial design alone.

References

Adams, M., Moore, G., Cox, T. J. and Croxford, B. 2007. The 24-hour city: residents' sensorial experiences, *Senses and Society*, 2: 201–216.

Appleyard, D., Lynch, K. and Meyer J. R. 1964. *The view from the road.* Cambridge, MA: MIT Press.

Bagnoli, A. 2009. Beyond the standard interview: the use of graphic elicitation and arts-based methods. *Qualitative Research* 9: 547–570.

Büscher, M., Urry J. and Witchger, K. (Eds) 2011. *Mobile methods.* London: Routledge.

Corner, J. and MacLean, A. S. 1996. *Taking measures across the American landscape.* Yale, NH: Yale University Press.

Cosgrove, D. and Daniels, S. (Eds) 1988. *The iconography of landscape.* Cambridge: Cambridge University Press.

Cresswell, T. 2006. *On the move.* London: Routledge.

de Certeau, M. 1984. Walking in the city. In *The practice of everyday life.* Berkeley: University of California Press.

Degen, M. and Rose, G. 2012. The sensory experiencing of urban design: the role of walking and perceptual memory. *Urban Studies*, 49: 3271–3287.

Department for Transport. 2009. *DfT Shared Space Project Stage 1: Appraisal of shared space.* London: DfT.

Dovey, K. 2010. *Becoming places: urbanism/architecture/identity/power.* London, Routledge.

Dulić, O. and Aladžić, V. 2016. *Architectural diagram of a city.* Proceedings of 3rd International Academic Conference on Places and Technologies 2016. Belgrade: University of Belgrade, Faculty of Architecture, pp. 85–92.

Farrelly, L. 2011. *Drawing for urban design.* London: Laurence King Publishing.

Fincham, B, McGuinness, M. and Murray, L. 2010. *Mobile methodologies.* Basingstoke: Palgrave Macmillan.

Garcia, M. (Ed.) 2010. *The diagrams of architecture.* Chichester: John Wiley & Sons.

Gehl Architects. 2010. Paving the way for city change. http://gehlarchitects.com/cases/new-road-brighton-uk/. Accessed 10/10/2016.

Goffman, E. 1966. *Behaviour in public spaces.* New York: Doubleday.

Green, D. 1985. A map of depravity. *Ten* 8: 36–43.

Howes, D. 2006. Charting the sensorial revolution. *Senses and Society* 1: 113–128.

Ingold, T. 2011. *Being alive: essays on movement, knowledge and description.* London: Routledge.

Jensen, O. B. 2013. *Staging mobilities.* Aldershot: Ashgate.

Jensen, O. B. 2014. *Designing mobilities.* Aalborg: Aalborg University Press.

Lefebvre, H. 1991. *The production of space* (translated by Nicholson-Smith, D.). Oxford: Blackwell.

Lefebvre, H. 2004. *Rhythmanalysis.* London: Continuum.

Lynch, K. 1960. *The image of the city.* Cambridge MA: MIT Press.

Lynch, K. 1981. *A theory of good city form.* Cambridge, MA: MIT Press.

Malnar, J. M. and Vodvarka, F. 2003. *Sensory design.* Minneapolis: University of Minnesota Press.

Marling, G. 2003. *Urban songlines.* Aalborg Universitetsforlag.

Merriman, P. 2006. A new look at the English landscape. *Cultural Geographies* 13: 78–105.

Murray, L. and Upstone, S. 2014. *Researching and representing mobilities: transdisciplinary encounters.* London: Palgrave.

Omidi, M. 2014. Anti-homeless spikes are just the latest in 'defensive urban architecture'. *The Guardian,* 12 June. http://www.theguardian.com/cities/2014/jun/12/anti-homeless-spikes-latest-defensive-urban-architecture. Accessed 10/10/2016.

Pallasmaa, J. 2005. *The eyes of the skin: architecture and the senses.* New York: John Wiley.

Pink, S. 2009. *Doing sensory ethnography.* London: Sage.

Priya Uteng, T. and Cresswell, T. (Eds) 2008. *Gendered mobilities.* London: Routledge.

Riis, J. 1890. *How the other half lives: studies among the tenements of New York.* New York: Charles Scribner's Sons.

Robertson, S. 2007. Visions of urban mobility: the Westway, London. *Cultural Geographies* 14: 74–91.

Rose, G. 2014. On the relation between 'visual research methods' and contemporary visual culture. *Sociological Review* 62: 24–46.

Sheller, M. and Urry, J. 2006. The new mobilities paradigm. *Environment and Planning A,* 38: 207–226.

Simmel, G. 1997 [1907]. The sociology of the senses. In *Simmel on culture.* Ed. F. Frisby and M. Featherstone. London: Sage, pp. 109–119.

Tilley, C. 1994. *A phenomenology of landscape.* Oxford: Berg.

Tschumi, B. 1981. *The Manhattan transcripts: theoretical projects.* New York and London: Academy Editions/St. Martin's Press.

Turnbull, A. 2013. *Exercise book.* Bexhill: De La Warr Pavilion.

Urry, J. 2007. *Mobilities.* London: Polity.

Vannini, P. 2012. *Ferry tales: mobility, place, and time on Canada's west coast.* New York: Routledge.

Vergunst, J. 2010. Rhythms of walking: history and presence in a city street. *Space and Culture* 13: 376–388.

Conclusion

Justin Spinney, Suzanne Reimer and Philip Pinch

Mobility is an arena in which design has exerted a strong but often overlooked influence; affording particular forms of mobility whilst limiting others. As designers seek to ensure the efficient movement of people and goods, so too they are implicated in the facilitation of social and civic goals (Knox 2010, 101), whether consciously or otherwise. As gatekeepers, designers both afford and limit alternative visions of and possibilities for movement. In *Mobilising design*, we have sought to tease out the intersections and connections between different registers, scales and categories of mobility in order to understand how particular places, identities and artefacts are made up. In this conclusion, we identify a range of themes that signpost both the collective contribution of chapters and avenues for further enquiry and exploration. We consider the effects of movement on design as unfinished accomplishments; the role of design techniques and practice in the process of conceiving the mobile subject; and the disciplining effects of design on the mobile subject. The final section highlights the importance of understanding designed objects and designers as moral and political actors.

Unfinished accomplishments

The first theme we wish to highlight is the idea of both mobility and design as unfinished accomplishments. It is all too easy to see ways of moving and design processes as fixed and unchanging. Chapters in this collection speak to a post-structuralist understanding of these practices as uncertain and becoming. Design is shown to be a process shaped by the boundary crossing and connective movements of practitioners and materials. In contributions from Thomas Birtchnell, John Urry and Justin Westgate; Kim Kullman; and Philip Pinch and Suzanne Reimer, design emerges as a practice that is 'unmoored', heterogeneous and much less emplaced than perhaps it once was. As indicated in chapters by Margo Annemans, Chantal Van Audenhove, Hilde Vermolen and Ann Heylighen; and Jayne Jeffries and Peter Wright, design practice can be resistant to change. Yet design practice is also transformed through movement, as we see in contributions from Kim Kullman and Craig Martin. Peter Cox's chapter demonstrates how meanings of design are changed as they are inserted into different worlds of sense-making and varying social and political contexts.

Similarly, the practices and subjects of mobility are seen to be shaped through more or less relational encounters between designers and users. As chapters by Anna Nikolaeva and Simon Cook demonstrate, spaces are conceived to discipline bodies to particular modes of comportment, yet the resulting spaces may semiotically define appropriate movement in ways that conflict with material design affordances in unintended ways. As Ole B. Jensen, Ditte Bendix Lanng and Simon Wind's chapter reveals, the relation between the lived and conceived becomes less the product of a linear and bounded process and more the product of constant flows.

Techniques and practices of design

Second, contributors to *Mobilising design* demonstrate a variety of design techniques and practices that are more or less open to embodying multiplicity, flexibility and "vagueness" (Miller 2006), in turn demonstrating different political orientations. Margo Annemans and colleagues; Emily Falconer; Lesley Murray and Susan Robertson; and Jayne Jeffries and Peter Wright all foreground the methodological and analytical underpinnings of design in capturing, analysing and representing mobility. What these authors emphasise is the varying distance created between lived and conceived through the use of particular techniques such as mobile ethnographies, drawing or video-recording. By asking how different design professionals understand others and become more or less porous to other ways of knowing and doing, these authors raise interesting questions regarding the politics of design in knowing 'the other'. By exploring the relationship between bodies and techniques, these authors collectively excavate the political implications of uncertain relationships where differing intensities of power accrue in some places and not others.

Calculation, discipline and embodiment

Foregrounding the moral and political dimensions of design invites us to engage with broader debates around how bodies are disciplined through design. Contemporary governmentality relies on the regulation of everyday conduct, and in particular modes of self-regulation that increasingly produce the individual as a calculating, self-interested actor (Ganti 2014, 95). Guy Julier engages with this debate in relation to design, arguing that designed objects such as the games console construct a micro-practice and calculative mode of being where progress is 'measured out'. Similarly, in their discussion of designing London's pedestrian sign system, Spencer Clark, Philip Pinch and Suzanne Reimer highlight the fact that the pedestrian is produced as a calculating and rational actor whether conceived of as a commuter or tourist; while Martin Emanuel's chapter considers the disciplining effect of traffic lights on both drivers and pedestrians in interwar Stockholm. In relation to car design, Lino Vital García-Verdugo discusses the closure of other ways of moving and relating due to the self-regarding focus of commercial vehicle designers. For Vital, the body and vehicle are positioned in relation to other bodies and spaces rather than in a unilinear relation to themselves. All of these accounts hint at the production of the calculating mobile subject by affording some things and proscribing others. Building upon this work, we argue that it would be helpful to connect themes of calculation, discipline and embodiment more explicitly in future work.

Morality and politics

Mobilising design has built upon understandings of designed objects and designers as moral and political actors (Latour 2002; Spinney *et al.* 2015; Verbeek 2011). As many chapters have attested, design is fundamentally a process of distilling and reducing: it is an inherently political process concerned with shaping what is and what is not important. Design filters and fixes a precise ideology out of a 'vague' set of everyday practices:

> representations by their nature are precise, and it is this act of precision that works against the vagueness and ambiguity of the world and, therefore, the openness of social life. It is in that movement from vagueness to precision where power relations are enacted.
>
> (Miller 2006, 464)

Within moments of transformation, power is enacted because it fixes identities and movements through defining what is appropriate. Which groups are allowed to define those aspects of the lived that are included in the conceived, therefore, can be highly problematic.

Accordingly, chapters in this book have built upon the ways in which mobility informs morality by design. Verbeek amongst others (2011, 21) has argued for the inclusion of artefacts and technologies in the shaping of everyday moral agency. Moral acts, he goes on to argue, depend on the ways in which human and nonhuman mediators are "integrally connected" (Verbeek 2011, 38). The contributors to *Mobilising design* have invited us to see moral agency as shaped by and distributed across socio-technical systems that encompass users, designers, mediums and objects. In so doing, the book argues that opening up design in relation to the production of mobile practices and subjects enables a more complete understanding of morality as the product of connections and disconnections. Seeing morality as an ongoing process of connection opens up a "research programme [...] focused on all the mechanisms and material mediations that make these [moral] affections, and the entanglements they reveal, visible and perceptible by the actors themselves" (Callon and Rabeharisoa 2004, 17). Here a designed object or project becomes more a "complex ecology" than a "static object" (Yaneva 2012, 93). Politics and morality are shaped by bringing mobility and design together.

References

Callon, M. and Rabeharisoa, V. 2004. Gino's lesson on humanity: genetics, mutual entanglements and the sociologist's role. *Economy and Society* 33: 1–27.
Ganti, T. 2014. Neoliberalism. *The Annual Review of Anthropology* 43: 89–104.

Knox, P.L. 2010. *Cities and design*. London: Routledge.

Latour, B. 2002. Morality and technology: the end of the means. *Theory Culture and Society* 19: 247–260.

Miller, V. 2006. The unmappable: vagueness and spatial experience. *Space and Culture* 9: 453–467.

Spinney, J., Kullman, K. and Golbuff, L. 2015. Driving the 'Starship Enterprise' through London: constructing the im/moral driver-citizen through HGV safety technology. *Geoforum* 64: 333–341.

Verbeek, P-P. 2011. *Moralizing technology: understanding and designing the morality of things*. Chicago: University of Chicago Press.

Yaneva, A. 2012. *Mapping controversies in architecture*. London: Routledge.

Index

Note: Page numbers in **bold** indicate figures or tables.